As my lecturer in 1978, Prof Cham was different from others as he took time to explain the practical aspects and applications of the subject matter. As Board Member and Chairman of our company since 2006, we benefitted from his analytical and systematic approach in guiding us.

After reading his autobiography, all the pieces seem to fall in place. This is a man with exceptional passion, talent and willpower. But most importantly, he has an ability to deliver, no matter how insurmountable the task. This was possibly one of the important attributes which several "Prominent Personalities" saw in him.

Prof Cham is my role model and I believe many of our younger generation, especially engineering students and young engineers, will find his autobiography inspirational.

***Wong Chi Keong**, B Engineering (Hons), University of Singapore*
Executive Director, Soup Restaurant Group Limited

The book contains Prof Cham's knowledge and experience in a wide range of fields as an educator, a banker, a portfolio investor, a corporate leader, a professional engineer and a transport expert. Reading his autobiography, I am enriched with a deeper knowledge about his extensive life experiences and network of famous personalities and friends. A riveting read.

***George P H Thia**, FCCA, FCA, MGer*
Formerly Merchant Banker with Morgan Grenfell Asia and
Merrill Lynch International Bank

The book captures the life journey of an extraordinary man which mirrors the development of Singapore from a third world to a first world country. I have benefitted from his stories. He was a key developer of human capital in Singapore as an educator and built NTU to be one of the top universities in the world. There is much wisdom to be distilled from his stories and from his involvement in the private and public sector. A truly remarkable Singapore boy who epitomises the spirit and contributions from our pioneer generation.

Tan Choon Seng
Ex CFO, MD Hewlett Packard, Group CEO Wearnes

This is a biography every Singaporean should read as it provides very rich insights on how the country has been shaped by the pioneer leaders in captivating anecdotal stories.

Professor Cham gives readers a good understanding of the country's challenges and progress through his sharp engineering mind. He has worked tirelessly as an educationalist and has guided companies and institutions as a member of multi-industrial boards and advisory councils. His contributions to the founding and building of NTU to one of the top universities recognised globally has enriched the nation. This is truly an insightful and inspiring book. Highly recommended.

Patrick Fang
Managing Director of PFANG & ASSOCIATES
Former Managing Director of Russell Reynolds Singapore

Although most autobiographies provide readers with great insights into the life and rich experiences of the author, this engaging autobiography not only succeeded in doing just that but also immerses the reader in a rich and inspiring narrative of the growth of Singapore for the past eight decades through the author's own immense contributions.

Professor Cham's contributions to Singapore span a rich spectrum of strategic areas in national development, be it in higher education management and infrastructure, corporate governance, industry transformation, defense, transportation, quality, media, social good, culture and many others.

Through his compelling and entertaining autobiography, he has given the reader a kaleidoscopic view of how he has helped to build a better Singapore. Another interesting feature of this autobiography are his thoughts on the future, the technological impact on the future of education, work and society. I have no doubt readers of any age and from anywhere, will find this autobiography a rich source of knowledge, in getting to know a prominent son of Singapore and in learning, reflecting, reminiscing, or understanding Singapore's transformation as well as into the foreseeable future.

Prof Lee Pui Mun
Professor, School of Business
Singapore University of Social Sciences

I am impressed by the range and depth of his contributions to Singapore over the 50 years of his active life. Prof Cham is well respected, especially for his insights into education, governance and businesses, and his vision on the way forward for education in light of Industry 4.0 with emphasis on re-skilling and up-skilling. He has consistently shown a unique grasp of problems, always sought to aid fellowmen in their hour of need. Prof Cham has gained the respect, esteem and affection of friends and colleagues. A must-read.

Dr Tan Lay Pheng*, Assistant Director*
Republic Polytechnic

About the Writer

Leong Ching is a former newspaper and television journalist and editor. She is now Associate Professor at the National University of Singapore and its Dean of Students. She has written books on companies such as Sembawang, Siemens and Cycle and Carriage.

CHAM TAO SOON: LIFE AT SPEED

Cham Tao Soon

Nanyang Technological University, Singapore

World Scientific

NEW JERSEY • LONDON • SINGAPORE • BEIJING • SHANGHAI • HONG KONG • TAIPEI • CHENNAI • TOKYO

Published by

World Scientific Publishing Co. Pte. Ltd.
5 Toh Tuck Link, Singapore 596224
USA office: 27 Warren Street, Suite 401-402, Hackensack, NJ 07601
UK office: 57 Shelton Street, Covent Garden, London WC2H 9HE

National Library Board, Singapore Cataloguing in Publication Data
Name(s): Cham, Tao Soon.
Title: Cham Tao Soon : life at speed / Cham Tao Soon.
Description: Singapore : World Scientific Publishing Co Pte Ltd, [2021]
Identifier(s): OCN 1179044028 | ISBN 978-981-122-708-0 (hardcover) |
ISBN 978-981-122-709-7 (ebook for institutions) | ISBN 978-981-122-710-3 (ebook for individuals)
Subject(s): LCSH: Cham, Tao Soon. | Nanyang Technological Institute--Presidents--Biography. |
Nanyang Technological University--Presidents--Biography. | College presidents--Singapore--Biography.
Classification: DDC 378.5957--dc23

British Library Cataloguing-in-Publication Data
A catalogue record for this book is available from the British Library.

For any available supplementary material, please visit
https://www.worldscientific.com/worldscibooks/10.1142/12012#t=suppl

Printed in Singapore

Preface

This autobiography serves as a record of my working life as well as my interactions with the various people in the course of my work.

I am glad that I have had the opportunity of a very fruitful working life. My working life started off in 1965 and lasted up till now. I have been kept very busy and I was very fortunate to be in some of the key industries and organisations like Keppel Corp, Wearnes, Natsteel, SPH, UOB, LTA, JTC, and SSO.

I had the opportunity of meeting prominent Singaporeans such as Tan Chin Tuan, Wee Cho Yaw as well as Mr Lee Kuan Yew, Dr Goh Keng Swee, Dr Toh Chin Chye, Mr Lim Kim San and SR Nathan.

I was very fortunate to meet and interact with them, which gave me not only a good perspective of various industries and businesses but also how some of these older statesmen think. In a way, it has moulded my own view of life.

For some time, I had been asked by my friends to consider writing my autobiography. I hesitated because I was not sure whether my stories are really of any interest to my readers. It was only recently, at the age of 80 and at the beginning of this decade, that I realised I need to put these thoughts on paper before my memory fades.

I am happy that I can still continue to contribute to society, doing whatever is necessary. I think this is my philosophy: "do what you can to make life interesting." I am writing this at age 80 and I still think I have a more useful life to come.

For this autography, I wish to thank Dr Leong Ching for help on writing and Dr KK Phua of World Scientific Publishing for getting the book done.

This book has been written almost entirely from memory, and memory is as inexact as engineering is precise. As far as possible, I have checked against historical documents; interpretations of events are my own, and I seek readers' understanding for any errors, which are entirely my responsibility.

Professor Cham Tao Soon
President (Emeritus)
Nanyang Technological University, Singapore
Jun 2020

Chronology According to Categories of Organisations

(a) Education

- NTU (1980–2002), NIE (1991–1998)
- US/NUS (1969–1983)
- Ngee Ann Tech College (1980–1982), Ngee Ann Polytechnic (1982–1990)
- NAFA (2003–2013)
- RI (2005–2013)
- SMU (1997–1999), SIM (1997–2018)
- UniSIM (2005–2014)
- LKYSPP, Board Member (2005–2012)
- Purcell School, London, Governor (1995–2000)
- Universitas Pelita Harapan, Indonesia (2000–present)

(b) Society

- Science Council (1975–1990), NSTB (1991–2003)
- Biomedical Research Council (2000–2002)
- Applied Research Corporation (1982–2000)
- National IT Committee (1992–1997)
- Advisory Council, Singapore Computer Society (1994–2003)
- Oxford & Cambridge Society, President (2001–2006)
- Rhodes Scholarship Selection Committee Chairman (1993–Jun 2005)
- Tan Chin Tuan Foundation (1999–present), Tecity Advisory (2003–present)

- Head Foundation, Chairman Advisory Board (2018–present)
- Singapore Symphony Orchestra (1999–2010)

(c) Engineering Institution

- Institution of Engineers Singapore (1971–1986), President (1980–1982)
- Professional Engineers Board (1973–1978)
- Academy of Engineering, Singapore (2011–present)

(d) Government

- Construction Industry Development Board, Chairman (1984–1991)
- JTC Board (1985–1995)
- SQA Governing Council (1993–present)
- Singapore International Foundation (2002–2013)
- Singapore China Foundation (2004–2014)
- Advisory Council on Community Relations on Defence (1984–1986)
- Singapore Design Council, TDB (1991–1995)
- Economic Planning Committee, MTI (1989–1991)
- Land Transport Authority (1999–2008)
- LTA Academy, Chairman (2008–2014)
- Singapore Rail Academy, SGRA (2016–present)
- Council of Presidential Advisers, CPA (2004–2010)

(e) Industry

- NSL (Natsteel) (1998–present)
- Singapore Press Holdings, Deputy Chairman (2004–2014)
- Keppel Corporation, Director (1982–2003), Keppel Technology Advisory Panel (KTAP), Chairman (2004–2011)
- UOB, Board (2001–2014)
- UOB (C.) (2004–present)
- Wearnes (1985–2011)

- MFS Tech, Chairman (2007–2011)
- Singapore Automotive Engineering, Chairman (1985–1991)
- SMRT (1987–1999)
- Soup Restaurant (2007–present)
- Advisory Board, Mott Macdonald Singapore (2004–2010)
- Advisory Board, SAP (APJ) (2014–2018)
- Asian Association of Business Incubation (2002–present)

Contents

Preface *vii*

Chronology According to Categories of Organisations *ix*

Chapter 1 My Family *1*

Family *1*

Chapter 2 My Education *7*

Primary School *7*

Goh Chok Tong *9*

Raffles Institution *12*

University of Malaya *20*

Cambridge *24*

Chapter 3 Educational Institutions *31*

University of Singapore *31*

A New Path for Engineering *36*

NTU/NTI *37*

Republic Polytechnic (RP) *56*

Nanyang Academy of Fine Arts (NAFA) *59*

Raffles Institution *61*

SMU/UniSIM *62*

Purcell School, London *71*

Chapter 4 Doing Good (Societies & NGOs) *73*

Rhodes Scholarship *73*

The Head Foundation *74*

Singapore Symphony Orchestra *76*

Chapter 5 Engineering Institutions *83*

Institution of Engineers, Singapore *83*

Professional Engineers Board (PEB) *86*

Academy of Engineering, Singapore *87*

Chapter 6 Government Bodies *91*

Science Council *91*

NSTB *93*

A*STAR *96*

Construction Industry Development Board (CIDB) *99*

Jurong Town Corporation (JTC) *100*

Singapore International Foundation (SIF) *101*

Singapore China Foundation (SCF) *102*

Economic Planning, 1991, MTI *104*

Land Transport Authority (LTA) *105*

Singapore Rail Academy (SGRA) *108*

Council of Presidential Advisors (CPA) *110*

Singapore Quality Awards (SQA) *112*

Chapter 7 Industries *115*

NSL (NatSteel, National Iron & Steel Mill) *115*

Singapore Press Holdings (SPH) *116*

Formula One (F1) *121*

Keppel Corporation *125*

Keppel Technology Advisory Panel (KTAP) *128*

UOB *130*

UOB (China) *131*

Wearnes Group (WBL) *132*

MFS Technology *133*

Singapore Automotive Engineering *134*

SMRT *136*

Soup Restaurant (SR) *139*

Mott MacDonald, Singapore *140*

SAP (APJ) *140*

Asian Association of Business Incubators (AABI) *140*

Chapter 8 Looking Forward *143*

The Impact of Automation and AI in Everyday Life *145*

Education in the Era of Disruptive Technology *151*

Chapter 9 Significant Personalities *161*

Mr Lee Kuan Yew (1988–1991) *161*

Dr Goh Keng Swee *164*

Dr Toh Chin Chye *166*

Tan Chin Tuan *167*

Mr Wee Cho Yaw *169*

Michael Fam *170*

Mr Tang I-Fang *173*

The Riady Family *174*

Mr SR Nathan *175*

Sim Kee Boon *177*

George Bogaars *178*

Lee Seng Tee *180*

Chapter 10 Reflections *181*

Technology and Impact on our Life *182*

Entertainment and Media *184*

National Crises *185*

Index *195*

Chapter 1

My Family

Family

My father was born in China and came to Singapore in his early 20s, while my mother was born and bred in Singapore. She had three children — two girls and one boy, I was in the middle between an older and younger sister. I had a half-brother who came with his mother from China after my father left China. For some time, we lived together in a small apartment in Tiong Bahru until the late 50s.

My mother was educated in Chinese while my father came to Singapore and had his English education at ACS. As a consequence, my father was bilingual.

My first memory was when I was 3 years old. I remember the Second World War — I remember the queues for our food and water supply. I remember that in front of our flat there was a vacant field, a small plot, where we planted tapioca and kang kong. We relied mainly on these for food, by boiling or steaming.

I also remember there was a Japanese soldier who visited our home — he wanted to have a drink of water. My mother served it to him. He got me to sit on his lap and I could smell the alcohol in his breath. All these, I can still remember. He did us no harm because my father was very smart — he had put a poster on the wall to say that we should respect the Japanese soldier, we should not confront them. The soldier probably saw that and did not harm us.

My mother used to be a school teacher, so during the war, she started a school within our house teaching Chinese. My father taught English because his ACS background enabled him to do so.

So, we had a vibrant school in the sitting room. Of course, we had to do it illegally, because the study of Chinese was forbidden during the Japanese Occupation. In fact, one of the outstanding students who came and studied Chinese from my mother was Peter Lim, who later became the editor-in-chief of the Straits Times.

We knew it was illegal. But what else could we have done? We needed money to survive during those times. We also sold small provisions by the roadside. We did everything we could to survive the Japanese Occupation.

My dad was non-political, but my mother was. As a Chinese school teacher, she was caught up in the milieu, to support Kuomintang just before the war, so she was anti-Japanese.

I still remember that for lunch, they would give us a tin can and some food. They would also give us coarse, poor quality rice in sacks. We did occasionally eat some rice, but we mainly had tapioca from our own plot. We somehow got palm oil, I don't know where my parents got this from but having palm oil was like finding treasure at the time. I know it is not very tasty, and it looks terribly yellow, but I was told that it kept us alive.

I am not sure whether there were amusements parks for poorer people like us. I knew about Great World, all the singing, the dancing, the merry-making. In our minds, it was meant for the tycoons and the business people, not us, what my parents would call the "very ordinary".

We heard about the Japanese surrender through word of mouth. So, things went back to normal, my father got back his government job, my mother joined a school to teach. The war was a momentous event for the world, the country and Singapore. But for a young boy, it was a memory of hunger, alcohol, fear, and the importance of family.

I initially met my wife through a common interest in music. I noticed a pretty and sweet little girl who very often seated quite near me. She had the grace of a lady. Later, a classmate of mine who knew her introduced her to me. After getting to know her, I discovered she had talent in more than music. She was strong in mathematics and very knowledgeable in science.

We had many common interests. She was studying in Convent of the Holy Infant Jesus (CHIJ) which was next to my school at Raffles Institution, and was one-year junior to me. When I had to go to KL for my university studies in 1960, we kept in touch through letters. When I came back for vacation, I dated her as often as possible. We became very close and knew we would be partners for life.

It was in early 1965 when I knew I was going to Cambridge for my PhD — so we decided to get married. In Singapore, we were married at the Registry of Marriage on 22 July 1965, and a few months later, 21 September 1965 in London according to the Chinese custom.

My wife Ee Lin came from a prestigious family — her grandfather Teo Siok Guan was the founder of Guan Hoe Ltd. She had 3 uncles and 8 aunties, 2 sisters and 1 brother. So, she had the experiences of a big family. On the other hand, my circle of family was small — so I had to learn from my wife's family tradition.

For most of her childhood, she lived with her grandmother in a big house at 1 Still Road. It was there that I made frequent visits to court her. From 1965, she lived with me in Cambridge and has been with me ever since. We had a son in 1969 and a daughter in 1975.

My wife has always been strong in mathematics and has a powerful sense of analytical reasoning. She graduated from University of Singapore in 1965 with Second Class Upper Honours in Physics.

When she accompanied me to Cambridge she worked as a Research Assistant at the Cavendish Laboratory, a famous Physics Laboratory.

She is also a good pianist, giving me much pleasure with her playing. We often have good discussions on classical music.

My son Tat Jen was born in 1969 in Singapore. He studied at Catholic Primary School, ACS and NJC. For his A Level examinations, he obtained straight Distinctions in the 4 Principal Subjects and 2 Scholarship Papers. As a result, he was successful in getting a Loke Cheng Kim Scholarship to go to Cambridge for his engineering undergraduate studies graduating with triple First Class Honours in 1993. He continued to pursue a PhD in computer engineering with a Research Fellowship from Jesus College, Cambridge which he completed in 1996.

On completion of his studies, he worked for Digital Computer in Boston which became Compaq and then HP. He returned to Singapore in 2011 and joined NTU as an Assoc Professor. Like his mother, Tat Jen is strong in Mathematics and has an analytical mind. He is now married and has 2 children.

My daughter Gee Len was born in 1975. She studied in Nanyang Primary School and then CHIJ. She is a very talented musician and got her LRSM in piano performance in 1991. One of the highlights of her piano achievements was the performance as a soloist in a Mozart Concerto with the Singapore Symphony Orchestra. Subsequently she was accepted by the Purcell School in London, which was set up for musically gifted children and obtained her GCSE A levels in Music and Mathematics.

She decided finally to major in Economics and enrolled herself in University College London and obtained her Economics Degree with First Class Honours in 1998. Subsequently she proceeded to MIT

Sloan School for her MBA in 2004. She has worked with Lehman Brothers, GIC, MAS, PWC(London) and SSO. She is the artist in the family, with an interest in photography as well.

My family life in 1977. I became a father in 1969, with the birth of my son Tat Jen. My daughter Gee Len was born seven years later.

Chapter 2

My Education

Primary School

I went to a Chinese school first because my mother was a Chinese school teacher. At that time, my home's cupboard was full of books — all Chinese. So, I was able to skip Primary 1 and went straight to Primary 2 in Gongshang at Outram. I studied there until 1949.

I was more advanced than the rest of the class, because my mother taught us at home. I was also exposed to a lot of riddles and difficult calculations. Gongshan was one of the top primary schools in Singapore, set up by a Hokkien association and was one of the most famous Hokkien Primary schools back then.

The medium of instruction — formally — was Chinese, but informally it was Hokkien. While the text and classes were taught in Mandarin, the explanations were in Hokkien. So, I picked up Hokkien very fast.

It is difficult to imagine today, but within the school, we had a zoo. In fact, that was the first time I learnt what a "porcupine" was. There was even a pangolin — as well as other very unusual animals. It was really a very good school and attracted many good students.

One of the students in my class was Lim Chin Siong. He was a very quiet person. Another person that is very well known in my class is Yeo Choo Kok, MP for Delta. At that time, he was three years older than I was. We ended up as colleagues at the Nanyang Technological

University. In the class, he was a leader, outspoken. In contrast to Chin Siong who hardly said anything and was very quiet.

Neither of us were among the top three students in our class — in a Chinese school, they honour first, second and third student. My ranking was not known, so it could not have been any good.

Because of the war, I was in school with people three years older than I was. When I was in Primary 5, the government at the time encouraged students to go to an English school. So, they created what they called a "special class".

My father encouraged me to join — "If you want to earn a living in Singapore, you better join an English school," he said.

There were only a few schools that had special classes that are meant to convert those from Chinese and Malay schools to English. The English special class was in Pasir Panjang English School. We did 2 years of "special class", then Primary 4 and Primary 5.

Other than language subjects, the rest was quite simple. At home, my mother gave us a lot of riddles and calculations to do — so it was not surprising that I could do these maths questions quite easily — sometimes more easily than the teachers after all, who had many subjects to teach.

At the end, we sat for entrance exam, now known as the PSLE. There were only 3 papers — English, Maths and Science. At the time, we were still a British colony.

We did two sessions of school — in the morning I did English, in the afternoon I went to Chinese school still. I was in Pasir Panjang English School special class 1 in 1950. In Gongshang I was in the last year, Primary 6. So, I did the Primary 6 in the afternoon. Then I got admitted into RI — I found out later that I was the fifth top student in Singapore at the time.

It was revealed by the Principal of Geylang English school who went to RI to teach. It had to be inside information as they did not release such information. The reason he revealed this information was because his student at Geylang had been the top student — he released the names of all top 10 students, and that was how I found out.

School was a happy time for me.

I was active in both sports and music. I played the recorder. For sports, I represented Pasir Panjang in table tennis, and later on, I represented Singapore in the sport. I was also in the school relay team. We had a lot of free time, no private tuition, so I just played. In fact, I have never had private tuition. At Tiong Bahru, there was a lot of opportunity for sports, especially at the community centre.

Goh Chok Tong

We were schoolmates in the Pasir Panjang Primary School from 1950 to 1953. He lived in Pasir Panjang and so it was convenient for him to go to school. On the other hand, I had to travel all the way from Tiong Bahru by bus because the school was the closest government school able to take Chinese school students into English schools.

Being older than Chok Tong, I was one year ahead of him.

We got to know each other well because of common interests. We were both in the scout movement and in the school swimming team. His speciality was the butterfly stroke and mine the breaststroke.

We went to RI for our secondary school education. Again, I got to know him well because we were both on the Prefectorial Board. In RI, the prefects were respected by the students because they were in charge of the discipline of the students.

We lost touch at the university level because he did Economics at the University of Singapore and I had to travel to Kuala Lumpur for the Engineering course. We did say hello to each other when we met at RI alumni functions or on the street.

One very memorable meeting was in July 1965. I had returned to Singapore because I was preparing to go to Cambridge for my PhD in September. Along Collyer Quay, I met him carrying a pile of books in his arms walking towards the Shell Tower (which has been demolished). I asked him what was happening and he told me he had resigned from his job at the Ministry of Finance to join Shell.

That was the end of our conversation because he was struggling with his books. A month later, I had to go to the Fullerton Building to sign my bond for my scholarship at Cambridge. The Ministry of Finance was also at the Fullerton Building at that time. I was surprised to see Chok Tong there. He told me Dr Goh Keng Swee instructed Shell not to take him and so he was back at the Ministry.

He left the Ministry to join Neptune Orient Lines in 1969 and I did meet him again a few times in the street. He told me that the job at NOL was not easy because shipping went through difficult times during that period.

The next meeting with him was also accidental. It was in 1976 just before the General Election. There was rumour that he was going to stand as a MP. I met him at the Singapore Polytechnic canteen at Prince Edward Road about a week before nomination. I asked him whether he would be standing for election, but he denied the rumour saying that there were other economists like Dr Goh Keng Swee and Hon Sui Sen. But when nomination day came, he was put up as a PAP candidate for Marine Parade.

The next occasion I met him was in 1979 when he was Minister for Trade and Industry. He chaired the CPTE Council for Professional and Technical Education and I was a member. We deliberated on the

Chok Tong and I at an SSO Dinner in 2009. We've known each other for 50 years, since we were schoolmates in Pasir Panjang Primary School in 1950.

need for technical manpower — arising from the meetings, NTI was created.

Over the years, we had several interactions. As Minister for Defence, I was on his Advisory Council on Community Relations on Defence in 1984. We worked on building good relationships between the military and the community. Then, when he was Prime Minister, he appointed me Chairman of Singapore China Foundation which was a Foundation building relationship between the two governments at the Civil Service level.

The Foundation was part of an agreement with Wen Jiabao, Premier of China. Later on, when he retired as a PM and became Chairman of MAS, we had good discussions on the management of UOB.

Chok Tong is an unassuming person, easy to work with. He is generally very open-minded, and he accepts everyone for who he is.

We have worked well in the various tasks. There were times he tried to interest me in politics, but he left me to decide my own future.

Raffles Institution

I spent the best part of my youth at RI (1954–1959). Although the school building was old, it had a charm of its own. When you walked along the corridor, the floor planks creaked. What made it unique were the students.

They had been chosen from the entrance exam from all government schools so it is no surprise that most were very bright. Surprisingly, many were also enthusiastic in sports and other curricular activities.

Form VA (1957) at Raffles Institution where I spent the best part of my youth. I am 4th from left, seated next to the form teacher. All prefects wore blazers.

Therefore, RI had all sorts of clubs, societies and interest groups. For each year, there were 5 classes of about 40 students each. There were also 3 classes for Pre-University I and II (those days, we called them Form VI year 1 & 2) one for Arts, one for Science and one for Medicine. For Secondary one (we called them Form II), the classes were named V, W, X, Y, Z.

The reason given was that the qualities of the students were placed in the classes at random. Only for the upper year, the classes were named A, B, C, D, E according to the ranking of positions at the end-of-year examinations.

The teachers were chosen among the pool of teachers for secondary schools by the Ministry of Education. From what I was told, the teachers at RI were above a certain minimum grade. Some were even former principals of primary schools.

I did find some of the teachers very knowledgeable in their subjects. For example, Mr Raymond D'Cruz was a good Mathematics teacher and knew the subject well. So was Mr Cheong Pak Lo for Chemistry. Mr Richard Tambyah was an inspirational teacher. He could inspire us not only to be good in our studies but also to be good leaders.

All in all, I think boys in RI were lucky to have an excellent group of teachers.

Since we had smart classmates, we learnt a lot from each other, maybe more than from our teachers. We discussed difficult problems together and found the best solutions. We not only worked hard, we also played hard together. Either we mingled in the same interest group or played together in some sports. RI is full of good memories for me.

In secondary school, I started to play the piano, and I took exams. Throughout my life, I did not have a music teacher, I was self-taught — you just need to observe, read, and learn.

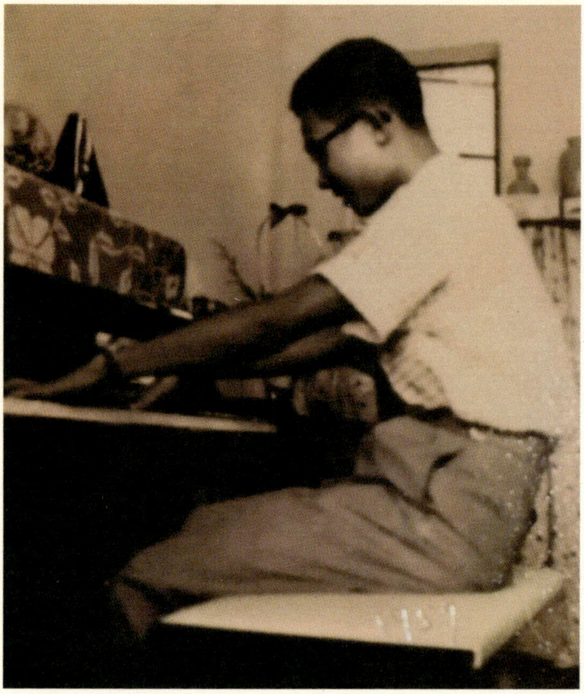

At my piano in 1958. I was never given formal lessons but I listened to my sisters playing the piano and taught myself.

My elder sister is a music teacher. My mother got a tutor to teach her to play piano. But she said her son cannot learn music. So, she never gave me the opportunity.

Still, I became a teacher to my sister — fingering is a matter of practice. I tell her about technique, about musical expression, the way to approach the music. At that time, the Raffles library, now the national library, had quite a few books on music — I read every one.

They were all on classical music. So, up till today, I do not know much about jazz or pop music. Pop, I do listen to a bit, some are quite nice but overall, it is not my taste. Now if you get in my car, I always have classical music on.

I played table tennis throughout my life. Because I played at the community centre, I represented the community centre at Tiong

Bahru. They came to notice me, and invited me to join the national squad. In Cambridge, I ended up as a table tennis champion.

I have competed only once against a Malaysian team. The Malaysian team came to Singapore, so I represented Singapore to play with the Malaysian team. Other than that, I don't think I took part in international competitions.

I played singles and doubles — I had a very good teammate from Tiong Bahru community centre. We played very well together. In fact, we ended up as Singapore 2nd, runner-up. Sadly, he has passed away.

I was the RI high jump champion for 3 consecutive years. My best was only 5 foot 6, not so high. We did not have the Fosbury Flop — that was only in 1968. High jumpers in my time landed in sand — if you do the flop, well, you won't be doing another jump any time soon.

In RI, I did participate in the orchestra. Besides piano, I also played a bit of violin, a bit of several wind instruments. Financially,

Receiving a book prize in 1957 for the best all-round student at Form V (Sec 4), reflecting both sporting and academic achievements.

these instruments were not too expensive — I got the cheapest violin. When I was in secondary school, the family was okay.

Music was not so popular at the time. It was only in the 70s then I supposed Singapore realised that society needed more than bread and butter. We had to have culture and entertainment to attract investments and visitors. In the 50s, 60s, music was just a hobby, it was the same to me also.

That was why when it came to university, my father said "No music, do engineering instead." That was how I ended up doing engineering. They were worried I spent too much time on music — which they considered as "wasted time".

My parents were practical people.

Thankfully, the teachers in RI were inspirational.

Teaching is a different thing from book learning. They talked about what it means to be a good citizen, how to do good for the world.

They taught us that a student should not aim to just pass exams but also to think about contributing to Singapore. We were fired up.

They spent the hour telling us about these things, and at the end of the lesson, sort of as a coda, they said "Oh you read page this to page that", and we did. That was how we learnt.

Of course, many teachers were very good, and taught us well. But in school work, we always had each other to fall back on — we could always consult the brighter student in any subject. I also did some teaching for my classmates. It was better than asking the teachers. So, nobody complained.

I appreciated our teachers for being so inspirational. The first Principal when we were there was Mr Ambiavagar and then after that John Young at my higher years. We were fearful of the Principal, because they expected discipline, very strict discipline. We were all very young, very playful.

The old compound is the present Raffles City, near to all the cinemas. There was Capitol, and on the other side was Odeon. The compound was very near to the sister schools, CHIJ at Victoria. The buildings which are still standing — such as CHIJ — are very old. I suppose you can say walking through them is like walking through a bit of history.

In RI, we had a very nice field in the middle, there is where all the activity, all the sports went on. I think what we liked most is that the compound was very central, we can go to almost any place we wanted to.

I supposed I had to be obedient because I was a Prefect. RI is one of the schools where the Prefectorial Board had a lot of power. Prefects were the ones who disciplined the students, not the teachers. So, prefects had to be well-behaved.

I was the table tennis team captain and a lot of students watched me do high jump so I was a well-known and popular guy. I was less known for table tennis. At sports day, everybody is there to watch you do the jump. For table tennis, the audience was quite thin.

As usual, in all the sports ACS was a competitor. Certain events, like high jump, Saint Joseph Institution had a particularly good high jumper, from what I remember. Overall ACS was the stronger team.

The rivalry with ACS was not academic because RI creamed off all the good students. With sports it was an even field. In fact, those who can study, on paper and in theory, they should not be so good in sports. But no, we were very strong in rugby. So, we always had competition with St Andrew. Athletics was with ACS but depending on the sport. Some sports like rugby match where all the students all gather and cheer, end up being very competitive.

We always went to the cinema, either Capitol or Odeon. Of course, being quite active in sports, I usually had to do training in the afternoon. Occasionally after class we just walk over to Capitol or Odeon — 1 dollar for a ticket, but it is still quite a lot of money in the 50s.

They showed mainly English movies. A lot are action, things like Tarzan, cowboys, when the hero is coming to the rescue, the whole theatre would clap. This hardly ever happens now — those were simpler times.

We did not skip school to watch a movie. But it was common for the boys to meet up with the girls from surrounding girls' school during social functions. So, sometimes, when Raffles Girls School had a function and invited some of the RI people, we would go over.

Similarly, we would invite some of the RGS and CHIJ girls. That was how I met my wife, from Convent. I met her in Pre-U days, because she was also very interested in music.

We graduated with Cambridge School Certificate, not O-levels.

Nowadays, parents are more pressurising. We had 8 subjects — Maths, Additional Maths, Chemistry, Physics, Geography, Chinese, English, English Literature. We went by the aggregate of the best 6 subjects — I had a total 7 points. In fact, 3 subjects I did not get 1 — Chinese, English, English Literature. I am rather weak on the arts.

It was not natural for most students to proceed to the Pre-U section in RI. In fact, some of my classmates who got very good results, decided to go and start work, for financial reasons. Some of them, the family needs them. In fact, I knew one who was a very good student and had to start working to support his family — in the end he ended up quite well in life anyway — as a senior staff of a Multi-National company.

I took science — with Maths, Further Maths, Chemistry, Physics. I did not take Bio, there is a story to it. When I was 15, we streamed into science. I took Biology then because my mother wanted me to become a doctor. Unfortunately, during the first week, we had to dissect a frog — take the frog's heart out while it was still pumping. I said forget it, I could not handle it.

I told my mother and I left the biology class. Of course, she was disappointed. That is why I never went into Medicine, although there was a medicine stream.

Soon after, my father decided my future for me. Those days, we were all very obedient and listened to our father. He was worried that I want to do music, so he asked, "What you want to do, what profession?" I asked what profession he wanted, and he replied, "Engineering, because you are strong in maths". I did not know what engineering was. But okay, I did engineering.

It was actually when I went into the first year in University that I understood what engineering was.

Medicine and Law were popular professions, engineering less so. From my class, only 5 of us went on to do Engineering. The rest did science — many of them later became teachers.

Pre-U was only 2 years. I continued with my games, my music, there was not much change. In fact, things got easier. Because I did not need to do Literature, Geography or Chinese. I got all 1s, all A.

So, I ended up as the top student in Singapore for Higher School Certificate (HSC). The other person with the same score as me was Tay Eng Soon — we were both top students in Singapore — 4 As. Dr Tay Eng Soon is 1 year younger than I am, because I had to repeat the 4 years in Chinese school.

Dr Tay Eng Soon and I both got the state scholarship, although we did not apply. Tay Eng Soon turned it down because he wanted to go overseas.

At that time, in 1959, PAP just came into power. Nationalists argued that our scholars should remain in Singapore. That is why Tony Tan stayed in Singapore doing Physics. I, for engineering had to go to Malaya. Tay Eng Soon refused to accept that. His parents being rich, sent him to England.

The state scholarship is equivalent to the present President Scholarship. It was called the state scholarship because we were then a state of Malaysia. It is only in 1965 that it became the President Scholarship, because we finally got a President.

My first reaction to receiving the state scholarship was, "Oh, money for the school fees!". We were practical. I got $3,000 on top of the fees, which was very good in 1960. In those days, a Mercedes only cost $8,000. So, I was very happy, in fact, the second year in the university, I actually bought a car. In return, I had to serve a 5-year bond.

University of Malaya

I applied to go to University of Malaya after I got the state scholarship. My parents were quite willing to let me leave — Kuala Lumpur is not too far away. The first time I set foot in Malaysia was just a year before in 1959. Part of this was because the Confrontation was still on, we did not know if it was safe in Malaysia.

There was also the Hock Lee bus strike. Many of my classmates were politically-minded — they questioned why we should be constrained by the British. Among my classmates, some went to jail after graduating — R Jody, G Rahman and Tan Jin Kwee. Quite a lot of people in my generation were riveted by politics and wanted to be part of it. I am afraid I am not very politically conscious at all.

I took a day train to KL — it was only after the second year that I decided to buy a car. After that I drove up, with some classmates from RI, SJI, ACS.

I went to the "second residential college" straight away during my first year. It was a University of Malaya hostel, but they called it "residential colleges". My college was right next to engineering — if I woke up late, I could just run down the slope to the class. There were 5 of us from Singapore — we got meals, lunch and dinner, even breakfast.

In the University of Malaya, Engineering was the first faculty to go to Pantai Valley, so everything was brand new and Engineering students were the only students there. We did not have any problems in getting accommodation because there was no competition.

The interesting part of my stay in second residential college was that Abdullah Badawi was actually my floormate. He was destined to be the future Prime Minister of Malaysia. He was there to do Religious and Malay studies. As a student, he was a quiet person. Originally, he had a roommate who was a Chinese, and I had a Malay roommate. After a while we exchanged roommates.

So that was the hostel life. I played games as usual — including table tennis. Other than that, I was mostly studying, because Engineering is a very tough course.

It took me 4 years to complete my course. 8 to 5 o'clock, every day including half days on Saturday. It is a very tough course, a lot of lectures, tutorials and practical sessions. So, there was not much time for anything else.

The University started in the University of Singapore — and they used trains to transport all the equipment to Pantai Valley. We all studied hard because we knew that casualty was very high. Many of the very good students took up to 6 years to graduate from the four-year course. Some took 8 years — because if you failed a year you had to repeat.

I have to say that the lecturers were not very helpful. I am afraid that the University, being new, did not know how all the equipment worked. Very often we had to figure out how to use the equipment ourselves. Unlike universities which have been established for many years and where students could take the notes from the previous years to help themselves, we did not have anything — we were the pioneers.

We had no specialisation, it was a broad course, civil, mechanical, electrical, we did everything. We called it "civil engineering", but then we did all sort of courses. The thinking at the time was that an engineer could end up in any one of the states of Malaysia, and may be one of the very few engineers in the state, so he must know a bit of everything.

That is why they thought we had to be much more broadly trained.

We had no access to computers. Only in 1965 when I went to Cambridge did I have a chance to get access to a computer. So, in the University of Malaya, my aim was to study, not much for socialization. After dinner, we would walk around the lake.

The campus is still there today, but of course it is very developed now. All the different disciplines are there and there are more halls of residences.

There were about 40 students per cohort that time. There were no ladies. The cohort was mostly Chinese, with a handful of Malay and Indians. Subsequently other faculties moved in. There were some social activities like dances and events. But except for a few of the more adventurous Engineering students, most ended up studying and studying.

I and 2 other Singaporeans managed to get first class honours which is very fortunate. In my cohort of 40, only about 10 of us finished it within 4 years.

I did study hard but I also studied out of the examination's syllabus. I was more curious about things in general. This is another fortunate characteristic about me.

In my first year, we had attachments. I had to spend 2 or 3 months in Keppel shipyard doing whatever they assigned me. The University sent us to our hometowns for attachments.

My time at the shipyard was not the most fruitful. We did not know what the purpose of the attachment was nor what we were supposed to do. So, we just went to a quiet corner, lay down and slept. Nobody cared because they were so busy. They thought we were a nuisance laying around.

In a way, it was true, we knew nothing in our first year, so they ignored us.

Nobody guided us, we just watched the technicians do their thing and when we were tired, we went to the corner. Keppel Shipyard had quite a lot of land, a lot of quiet corners to just lay down and have a good nap.

In my second year I managed to get an attachment in developing a Shell refinery in Port Dickson. Because they had just started at that time, they were building up the facilities, so I was part of the Engineering team.

In my third year, I was back in Singapore attached to the public works department doing some rather meaningful work. By that time, I was almost an engineer. Engineering did not really interest me, because civil engineering is a lot of common sense. I wanted something exciting and challenging.

In civil engineering maybe 60–70% of the subjects are common sense subjects. Of course, there were some challenging parts, 30% where some rigour is needed and that is where I excelled — I studied ahead and spent quite a lot of time on them. Sometimes, I helped out the lecturers who got stuck at a problem on the blackboard.

The only person who came to my convocation in KL was my future wife Teo Ee Lin.

In my year, there were 3 of us who got first class honour out of the 40 who graduated. All 3 happened to be Singaporeans. There were 2 who got second class upper.

Only one person came to my convocation — my future wife. My parents never came, they thought it was too much trouble coming to KL.

I told the Singapore Government, I wanted to further my studies, so they agreed to postpone my 5 years contract. I applied and got the Commonwealth scholarship to go Cambridge to do Aeronautic studies. For which I got slapped on another 5 years, so my bond was 10 years.

Cambridge

Cambridge was the most advanced university in aerodynamics. Because I already knew the subject quite well, when I combed through Universities, I was sure that Cambridge was the place.

It was not easy to get into Cambridge. Before going there, I wrote a scientific paper on one aspect of aerodynamics, which I sent it to Cambridge. The professors were impressed, and invited me to Cambridge. After that, I applied for the scholarship.

I did not know how to prepare for going somewhere so far away — it was the first time I got on an aeroplane in 1965. During those days the plane was so slow — it took more than 24 hours to reach London. I think that time they called it, BOAC, British Overseas Air Corporation.

It was really a cultural shock because I never been to a place like England. I still remember taking a shower — I did not know how to turn the hot water on. So my first shower was all cold water — that was in September — quite cold!

But that time I was very ignorant — only later did I ask for help and learned.

I did not see any drains. In Singapore, we have big drains. That was the first thing that caught me by surprise. In London, the drains are all underground.

At that time, we were gathered in London because there was an orientation for the scholars from all over the world. British Council made the arrangement — it was a Commonwealth scholarship. So we were there, and of course, my future wife came with me. But then she had relatives in London so she later went to stay with her grandmother.

They brought us to all sort of places, even to the palace to meet the Queen. Because we were Commonwealth scholars, they wanted to build good relationships. After a week of orientation, we went to Cambridge. Because I was going to be married, I stayed together with a family. I rented two rooms, the sitting room and the bedroom and a small kitchen.

After we were married in London, my wife came and joined me in Cambridge.

We had registered our marriage, but did not have a proper ceremony. Her family being old-fashioned wanted to do it in London. She has lots of relatives in London, so we had a dinner ceremony.

I was supposed to work with Professor Brooke Benjamin, but I found it quite difficult. He slept in the daytime and worked in the night so nobody would disturb him. He would make appointments to see me about 10 o'clock in the morning, but he was yawning half the time.

Formal dinner at St Catharine's College, where I studied Engineering.

I discovered that he was a somewhat odd professor — he was very into his research and a well-known eccentric. I looked at his work schedule and said "No way."

I decided I could not work with a professor whom I cannot see. I switched supervisor and worked on another topic. Of course, I remember up to today — that he was instrumental in getting me to Cambridge.

In Cambridge, I went to engineering and my wife, as a graduate in Physics, worked as a research assistant in a Cavendish Lab.

I was at St Catharine's College. To finish a PhD, you needed 9 terms. Not quite 3 years. But by the end of second year, I had already finished all my work. So I had time to kill.

I was interested in the turbulent layer, as applied to a rotating body so one of the applications was on rotating missiles. My supervisor was an expert in that, so I worked with him to come up with certain observations and solutions on how to predict models — so we ended up with quite a number of papers in scholarly journals.

Within 2 years, I finished my thesis. My supervisor said, "You cannot earn a degree unless you finished the 9th term."

So he just assigned me some things to do, more for a research contract rather than my thesis. I had to wait until end of the 9th term before I could submit my thesis, which was about May of the following year. He paid me for this research work, so it was okay.

Every break, my wife and I would go to almost every neighbouring country. That was the opportunity to see Europe. We also travelled quite a lot of UK, we went to Scotland, Wales. During the break I did not have to work on my thesis, so we could enjoy ourselves and travel.

I graduated with a PhD from Cambridge in 1968 — three years after I enrolled.

I was not used to the British way of life. But one thing I liked about Cambridge was the environment, so different from Singapore. It really is a place of learning. If you ask me, where to go if you had a scholarship, I would say go to Cambridge. When I was studying there,

I would go to the lab or the computer centre and there were professors in their 70s still working beside me. Everybody is so keen on learning.

I had the opportunity to attend lectures by very eminent professors for free, anybody could go to any class. They do not chase you out. I even managed to sit in the class of Paul Dirac, a very famous physicist. There were so many of them — I nearly had the chance to go to a lecture by Stephen Hawking, but unfortunately, it clashed with another class on my timetable.

My research topic was half theorical, half experimental. For the theoretical part, I used the most powerful computer in the world at that time — called Titan. Even then, I was given memory of only 50k words (one word was 8 bytes, but nobody used "bytes" then). Nowadays, even my phone has "megabytes".

I had to submit the programme by 10 pm, and could only get the results the next morning because the machine took a long time to "churn" out the results — this was even before punch card, it was on the tape.

The tape was one whole roll — if I made a mistake, I would have to start all over again. It was very tedious, I had to be very careful to make sure the punch tape was done properly.

We would submit the tape at night. The next morning, we would get our result. I looked at the timing — it took half an hour to run my programme. If I had to make changes, I would have to start again then wait for the night before I could send the tape in and wait overnight before I got the result. It shows that the most powerful computer in the world at that time, was very elementary when compared to today.

Titan was the mainframe. There were only 2 such powerful computers in the world, both owned by the British government, which let Manchester and Cambridge to use one each.

There is a big difference between the facilities in Cambridge and Malaya — a different scale of sophistication and opportunity — there was no way for Singapore to be at the front end of research at that time.

That is why when I returned to Singapore in 1969, I decided that there was no point doing research because we did not have the environment, or the equipment. We also did not have the will for doing high-level research. At our stage of development, there were more urgent tasks.

Chapter 3

Educational Institutions

University of Singapore

After we became independent in 1965, the University of Singapore separated from University of Malaya. That is when we knew we must start our own Faculty of Engineering. In terms of timing, it was perfect for me as I had just ended my PhD.

I liked Cambridge very much but I thought of coming home and contributing to Singapore.

For the interim period between 1965 and 1968, Singaporeans who wanted to study Engineering had to go to Kuala Lumpur to the University of Malaya. The Faculty of Engineering in Singapore was officially slated to start only on the first of April 1969. I remembered that people were jokingly asking whether it was an April Fool's joke.

At our official opening, we already had fourth-year students.

That was because the University contracted out the first three years of the engineering degree to the polytechnics. We also recognised polytechnic graduates as equivalent to third-year students. As they only needed another year to get a degree, many of them stayed on at school, even when they had already started working — they left their jobs which was a difficult decision in those lean times.

We were initially located at the Singapore Polytechnic because that was where the equipment was.

Although we were not a formal faculty, we already had students. We had special labs just for the degree programme although it was part of the polytechnic. This caused some friction between the lecturers that were teaching the polytechnic students and the lecturers teaching the university students because we shared the same campus.

Among the first batch of graduate students were some well-known names. In the 1969 batch of graduates, we had Chew Choon Seng who ended up as SIA's CEO and chairman of stock exchange. Another notable graduate is Kwa Chong Seng, Mobil chairman and SGX chairman.

Initially, we had a very small intake of 60 for Year One. But we grew quickly as it was part of the government initiative to produce more engineers. When I was the Dean of Engineering in 1978, our intake was 300 every year. The Prime Minister's Office directed us to increase this to 1,200. The very next year, we started taking in 600, double the previous intake. The aim was that half of all university graduates should be in engineering. Other faculties such as Science would take fewer students.

In 1976, the Engineering faculty was the first to move into Kent Ridge campus. By that time, we had been cramped at Singapore Polytechnic for 7 years — everybody looked forward to moving.

We had some hostels ready when we moved in — but supply soon ran short as students from architecture moved in. At the time, many students did not live on campus — they felt that Singapore was not that large so they could make do with a bit of travelling.

We were growing very rapidly and one of the difficulties was in recruiting. We knew that in the long run, the stability of the faculty depended on Singaporeans. I wanted to have about 60 to 70% Singaporean staff, and the rest expatriate staff who come and go as it is good to have renewal of academic interest.

We did not give permanent contract to expatriate staff. We attracted good expatriates and they were very mobile — they would come here for 3 years to find out more about Singapore and the region. Then, they would not hesitate to move away to Australia or Hong Kong.

There were not many Singaporeans doing Engineering then. Because of the industrialisation programme in the early 70s, the government was very concerned about producing enough engineers. PSC only gave engineering scholarships. If you did well, it was a choice of either taking engineering with scholarship or no scholarship at all.

As a result, in the mid-70s, many permanent secretaries were engineers by training. Most of the ministers were engineers by training too. Unfortunately, they do not practice. So, it is not true that we don't have engineers — in fact, we had a lot of engineers but they end up in the administrative service, or being roped in as politicians.

When we expanded to 1,200 students, there were many engineers in the work force.

Here is a story about how I get some of the brainiest Singaporeans to join the faculty of engineering.

I started a senior tutorship scheme then in the University of Singapore. It happened while Dr Lee Wei Ling was graduating as a doctor in the University of Singapore. At the convocation in the Trade Union House in 1979, Mr and Mrs Lee Kuan Yew turned up for the convocation.

There was always a tea at the end of convocation. Mr and Mrs Lee were in a separate room with a few senior people. As one would expect, nobody dared to stand near them. Seeing them alone, I plucked out my courage and talked to Mr Lee.

He knew I was the Dean of Engineering so he asked me if there were any problems in the Faculty of Engineering that we could improve on. I said yes: "It is unfortunate that all of our scholars were made to join the administrative service by the Public Service Commission. If we did not have good graduates come into the faculty, how could we ever be good?"

He saw my point at once and asked "Can you attract them?" I said yes as I have a senior tutorship scheme that can perhaps attract them. In the scheme, all returning scholars, have a choice of joining the admin service or taking up our senior tutorship scheme where they will do some teaching.

They will then complete their Master degrees at the University of Singapore as we wanted them to do some teaching here. At the end of the Masters, we would send them for scholarship to any place in the world to do their PhD. While they are on senior tutorship and overseas, we would pay them their salary as part of their scholarship.

It was an expensive scheme but I told PM that it was the only way to attract good students. I had proposed the scheme to the Ministry of Finance but as expected, they rejected it as it was too expensive. I told Mr Lee that we would never be able to elevate the Faculty of Engineering otherwise.

So, after the convocation, he overturned the decision of the Ministry of Finance.

The senior tutorship scheme took effect from 1980 onwards. We managed to attract 100 scholars to join both the Faculty of Engineering and Faculty of Science. Many of them thanked me for speaking up. In fact, one person who was on this tutorship scheme, is now the President of NUS Professor Tan Eng Chye.

Many of NUS star professors actually started out as senior tutors. The entire process took a long time of about 2 years followed by another 3 years, and another 5 years. In the long run, many of the key people in engineering came from this senior tutorship scheme. It is something which I am very happy to achieve.

When I was the Dean, only 10% of the students were female. The females were younger than the males, because the males had to serve National Service. The difference was obvious.

It was a very popular faculty then because of the industrialisation programme and the government promised that if you wanted a stable and good paying job, you had to be an engineer. We attracted very good students to Engineering. Later this was not so easy, as other professions became attractive too.

We liked to have a balanced of genders so we got some of the successful engineers such as Ho Ching to come and give talks about women in engineering. As we progressed, more IT came into engineering to deal with computers.

When we started taking 600 students, we had 200 in each of the civil, electrical and mechanical engineering disciplines. Civil engineering was the most popular course as they study infrastructure and development which needed a lot of engineers.

Civil engineers had a good time and that attracted better students. It was only later when industrialisation became successful in the 80s, that mechanical and electrical engineering became popular.

I felt strongly that students should be attached to industries. It was hard work to find placements for them. Fortunately, our industries and building construction works expanded, so we never had any problems giving them meaningful attachments.

In fact, I did a lot of travelling because we had to get more staff including expatriates. I had to go travel all over the place to recruit them. US, UK and Australia are popular places. I remember going to Sri Lanka at least three times. They had a political problem between the Tamils and the Sinhalese — we were able to bring in quite a few professors. Many of them are still here in Singapore.

Fortunately, our government realised that if we wanted to get good staff especially from the West, we had to be competitive so our salary scheme became quite attractive. I had no problem in attracting good ones from US and UK because our salary was always higher than the UK.

In addition to travelling overseas, we made use of professors from the US and the UK, who knew us and understood what we needed, to help us with interviews.

A New Path for Engineering

As the Dean, I participated in National Manpower Planning, headed by the Ministry of Trade and Industry. In 1979, I was involved in the CPTE or Council for Professional and Technical Education. The council was formed because the government felt that we were at a new stage of economic policies and we no longer needed to depend on cheap labour or substitution of import.

We wanted to go to the next level of value-adding and attracting businesses especially in high tech. The economic policy also advocated higher wages. In order to justify that, workers needed to provide value to the economy.

We needed not just engineers, but also technicians. The aim of the CPTE was for manpower planning — to get the University and the polytechnic to produce graduates who can bring Singapore to the next level.

A project like this in other countries would have the Minister of Education as the Chairman, but in our case, the Chairman was the

Minister of the Trade Industry. The transformation was not for education's sake but for industrialisation.

The Ministry of Education and Ministry of Finance was there when we came to an agreement, together with NTUC and Lim Chong Yah, Chairman of the National Wage Council.

We spent a lot of time debating what sort of engineers should be produced. We came to conclusion that only a quarter should be research — three quarters should be practice-oriented. I created the term, "practice-orientated engineers". In order to do so, we had to create a new institute, which we did not yet call a university.

We needed to have 900 places for practice-oriented engineers. The planning committee had to prepare the ground to start this institute.

NTU/NTI

Nanyang Technological University (NTU) started on the foundations of the old Nanyang University, the first university built by overseas Chinese.

Under the difficult circumstances who saw the closure of Nantah, Mr Lee Kuan Yew promised Singaporeans 3 things in order to for it to close. First, we will receive a better university for one that was taken away. Second, we will use the Nanyang University campus and retain the name Nanyang.

Third, it would only become a University in 10 years' time. His rationale was that we should not close a university today only to start a new university tomorrow.

As if true to the timing while the Nanyang Technological Institute started in 1981, and in 1991, we became a University. (At first, we wanted it be named Nanyang Institute of Technology, but the acronym was NIT so when we turn it around — Nanyang Technology Institute or NTI).

Tony Tan was very interested in the development of NTU, because the closing of Nanyang University and the starting of NTU was not just an academic issue, but also a political one.

Often, when we discuss complex matters on the phone, he gave the go ahead after only 5 mins.

I must say to some extent I did not observe good governance because for NTI or NTU — we had a board of governors, but there were times when I informed them of decisions only after I had decided to do it. Under proper corporate governance I should have gotten their approval — but the trouble is that it takes time for it to be approved and by the time approval is received, it is often no longer useful.

Dr Tan's personal interest and attention to the development of NTU was the main reason things happened so fast. Our first round of requests for the physical development of NTU which we called the "North Spine" was for $180 million.

When the NTI project was announced, it was actually the Ministry of Finance who asked me and the Registrar for the papers which we were to have submitted for funding.

At that time, the Registrar was the COO. He was not only looking at the registry but also HR (and everything else). When he received the request from Ministry of Finance, he asked me how we were going to do that. I said "No problem, you write up a paper and form a rough estimate."

Having been the Dean of Engineering, I had been looking after many students. I knew that we were designing a place to hold 3,000 students and could estimate what was needed to support them.

"Just put a rough figure down. The Ministry of Finance will typically deliberate on it for 6 months before we get their decision. In the meantime, we would have time to plan it."

As it turned out, we had no time to plan.

We submitted it and lo and behold — it was approved one week later.

We were not clear actually what we were approved to do, so we started getting the architects and planners to say that we got the $180 million and asked if we could build according to what is available.

On 1982, the front page of the Straits Times announced "Three-Finger home for Nanyang: NTI unveils $176m complex."

The iconic campus was designed by Kenzo Tange a well-known architect planner. He was *en route* to Singapore and he offered his service for free. It was an offer we could not refuse and so invited him to come up with a proposal.

NTU was designed by renowned Japanese architect Kenzo Tange (centre).

He came up with a proposal using topography, unlike other architects who planned the building on paper, so, we had to build the model of the campus and the terrain on site.

By the time he finished, we had 2 proposals. One follows the planning parameter of NUS which meant that we hugged the level of the hills. A second bolder suggestion was to have bridges across NTU. He presented them to a committee appointed by the government comprising Ong Teng Cheong who was an architect, Teh Cheang Wan (then Minister for National Development) and Tony Tan.

They went for the bold proposal, which still forms the heart of NTU today.

I participated in the Council for Professional and Technical Education (CPTE) which concluded that we needed to start the new institute which was to become NTI. I knew I was considered for the role of the President of NTI, but I was quite reluctant for 2 reasons. First, I had to spend a lot of time on it and it was during the time in my life when my children were growing up. Second, my wife was also quite busy working.

In the end, I offered to take on two positions for the next two years — the Dean of Engineering in NUS and the President of NTI. The reason is, so that if I faced too much opposition from the Nanyang University Alumni for the plans I had in mind, I could go back to Engineering.

It was only in 1983 that I told Lim Pin, Vice Chancellor of NUS that I could manage NTI and so resigned as a Dean to focus full-time on NTI.

I held a lot of meetings with employers to get them to accept our training. At one point, NTU graduates were doing better than NUS graduates. Second, NTU managed to get good students. In fact, I tracked student admissions from our first year — three-quarter of students came to NTU and a quarter remained in NUS.

Of the quarter of all students remaining in NUS, I tracked their scores (I could still track because I was still the Dean of Engineering) — this 25% get a choice of remaining in NUS or NTI. I was relieved in 1982 when half of them opted to go to NTI. We had good inaugural graduates like Lee Bee Wah and Inderjit Singh, both who went on to be Members of Parliament.

What Singapore needed at the time was practising engineers so that our programme was different from NUS. We had made sure that we had a great deal of publicity around the narrative that the professors of NTI had track records as practitioners.

In NUS, 90% of the professors have never worked a single day in the industry. So, we made sure to underline this fact so that students had confidence that NTI had a good track because we were practice-oriented. I had innovative practices such as 6-month long industrial attachment in the third year which NUS had never implemented.

Many of the projects in NTI were done by students. They really practised as engineers. As this was a largely practical programme, we were surprised that in 1985 we were named as the best engineering programme in the world by the Commonwealth Engineering Council. So, our model was validated.

We were also validated by the top people we managed to attract. Professor Bengt Broms was a Swede and a very prominent professor in soil mechanics. For some reason, he was interested in NTI in 1982 to 1983 when we had barely started. He joined Nanyang Technological Institute in Singapore in 1983, where he was Professor and Head of the Division of Geotechnics and Surveying until his retirement at the end of 1995. He had attracted many prominent professors to his division.

The Nantah Heritage

Even as we were making changes at breakneck speed, I ensured that we never lost sight of our strong heritage. We preserved the formal Nanyang University landmark — the arch at the old entrance — it is still there today.

Chairman of the former Nanyang University Council Wee Cho Yaw (left) with Minister George Yeo (then at Ministry of Information and the Arts) and myself, along Yunnan Gardens in 1995. The Nantah landmark — the arch of the old entrance, still stands there today.

We put up some plants so that it looked like a mini-park for the residents in that area. You will notice that we have a building that has a distinctive Chinese design. It was used as a library and is now the Chinese Heritage Centre. We preserved the Yunnan Gardens, a few old pagodas and the Nantah Lake.

Our new campus worked around the old elements. While Kenzo Tange did the master concept, the details were done by local architects.

Kumagai Gumi, a Japanese firm, tendered for works on the North Spine. It was not an easy job — I am glad that it was Kumagai Gumi who won the contract as they were disciplined with housekeeping (a problem for many construction sites) and were good contractors. We used a number of the old hostels and renovated them completely in the 80s. Many were very old and left behind from the 50s.

When the new campus at North Spine was ready in 1986, we had an official opening. We already had our first batch of graduates and so it was a grand celebration. President Wee Kim Wee was Guest of Honour, Prime Minister Lee Kuan Yew and many of the Nanyang University Council members also attended.

We had succeeded despite the odds — a difficult political start, a punishing schedule, and the long shadow of an older and more successful university.

I think there are 2 reasons why NTU did well. One, our aggressive marketing of our engineering degree as a practice-oriented course which makes successful engineers. For the students, it did not matter that it was something new as they could get NUS degrees. We told our NTI students that having practice-oriented education was the best way.

The second is that many of those starting the university with me actually felt that it was a good challenge. We had a good programme as we recruited lecturers that had real engineering experience. We earned our mark as one of the best universities in the world in engineering by 1985 — when we had our first batch of graduates.

There were 2 teething problems for NTU — first it is far from the city and the journey to NTU takes a long time. Second, the students were overwhelming male.

The planning and development committee decided that once our engineering discipline was established, we would complement it with other disciplines. In 1986, we started to plan for a business school. In Singapore at the time, people respected professions rather than business schools or business administration. It is not a *profession,* compared to say, accountants.

When Tony Tan came to know about it, he was against the idea of us starting an Accountancy school to compete with NUS — instead, he wanted us to transfer accountancy from NUS to NTI.

However, the accountancy students were not happy as they felt NTI was too far.

In order to accommodate them, we took over some HDB flats across AYE and took some trouble to do them up nicely. Michael Fam was the chairman of HDB and chairman of NTI then. Our lease was for five years, so all the accountancy students could stay in nice flats. We even arranged transport to fetch them to NTI every morning.

By 1991, I thought that we had done quite well and were as good as other universities in the world. We decided to ask the government if we could award our own degrees. They agreed — so instead of NTI, we were now NTU.

Mr Michael Fam was a hands-on chairman for the Nanyang Technological Institute, and it was due in large part to his indefatigable work ethics that we kept to schedule.

In celebrating changing our status, the Guest of Honour was Mr Goh Chok Tong then Prime Minister. One of the three conditions laid down by Lee Kuan Yew to convince the Nanyang University Council to accept the closure and merge with University of Singapore was to retain the name "Nanyang".

We kept the promise. There was some concern that people may not know where "Nanyang" was (unlike the National University of Singapore, whose geography was identified immediately), so we had thought about calling ourselves NTUS, Nanyang Technological University of Singapore. But we picked a simpler name. Over time NTU, rich in history and strong in mission, found its place in the world.

Building NTU

There were 3 pieces of advice from colleagues in England which have helped me in shaping NTI and NTU.

The first one is Professor John Brown, who was then head of the electrical engineering department of Imperial college and the President of the Institution of Engineers in the UK at one point in time. He came to advise me on how to shape a good practice-oriented course. As President, he understood what practice-oriented courses meant. He also came to Singapore and helped us sell to our employers, potential students and parents.

He made the strong case that in the 1980s, Singapore required more practice-oriented engineers rather than research-oriented ones as it was developing quickly. He was one of the first to support NTI and rallied support from various other quarters.

We were grateful to him and awarded him an honorary degree from NTU.

Sir Eric Ash also contributed to the second half of NTU's development. He was the Rector of Imperial College, and he spent a few weeks in NTU to see where we should improve our teaching, research and how do we deal with the industry, especially the high-tech ones. He came up with a report towards the end of the 90s.

As I had come from Cambridge engineering, I knew the Engineering Dean well. We came to an agreement that they could help us with research. Each year, we had 3 professors from the engineering department to spend one to three months with us to help us. We signed a 5-years agreement and rewarded the visiting professors quite handsomely.

When Imperial College found out, they too wanted to join and signed up. So, for five years, we had three professors coming from Cambridge engineering and three professors from Imperial College. They all helped us with our research in NTU.

Lord Dainton had written the report arguing for the closing of Nantah. (Lord and Mrs Dainton in my office in 1990).

The third man was Lord Frederick Dainton. I remember him very well — he was instrumental in the Government's decision to close Nanyang University. In 1979, he wrote the Dainton Report which pointed out that "the arguments for maintaining two universities [University of Singapore and Nanyang University] whether on two campuses or one (an absurd proposition) are extremely weak whilst those in favour of a single, strong university at Kent Ridge are compelling and I recommend accordingly".

Later, he also told Tony Tan that NTI was ready to confer its own degrees, and to be a comprehensive university with strength in other disciplines as well as developing our research.

We became good friends and as an Honorary Fellow of St Catharine's in Cambridge, he recommended me as an Honorary Fellow too. So that was how I become an Honorary Fellow in my own College.

Practice and Engineering

We did internships from Day 1 of 1982 at NTI. In fact, this was one of the plus points which helped us to be ranked as one of the top engineering universities in the world.

We had an industry liaison unit (LIU). In fact, that particular staff member had remained with me throughout my period in NTI/NTU because he found that the work was so satisfying and the relationship with our external partners was good.

During my days at both NTI/NTU, funding came directly from the Ministry of Finance. I had to deal with the Ministry of Finance rather than the Ministry of Education. I met a number of permanent secretaries from the Ministry of Finance.

There were some incidents that I thought would be good to pen it down.

The first one was with George Bogaars as a PS for Ministry of Finance. At that time, he was working under Minister Hon Sui Sen. I had suggested that NTI should be put on par with NUS. I explained that NUS had associate professors and professors at the top band then. NTI just had principal lecturers and senior lecturers. I felt that unless we were seen to be on par with universities, we could not attract good staff.

The perm sec disagreed — he said that NUS is a university and NTI was *not* a university.

I went to Tony Tan and complained that we would be handicapped before we started. Tony Tan talked to Hon Sui Sen but to no avail. In the end, Dr Tan resorted to asking Lee Kuan Yew who as PM, overruled the Ministry of Finance.

That was how we became equal to NUS. In term of costs, NTI's budget went considerably higher because the salary of these professors were two grades higher.

In the early years of NTI, the administration building leading to the library was completed but was a bit bare. Being concrete, it could be quite hot in the afternoons. One Sunday afternoon, Mr Lee Kuan Yew came without advance notice (as usual).

He must have felt very hot. The next day, on Monday, I got a note from him to saying that we must try to cover the concrete part of the North Spine. He had been to Paris, and saw a very nice structure called a "cloud structure" — a canopy that could be used to shelter the concrete area.

He sent me the photograph and suggested we did the same for NTI. We traced the project architect and found that he was in Australia. We got him to come over, do a sketch of what should be done to our area and to give an estimated cost. It was quite costly — I had to put up to the Ministry of Finance for money. The perm sec, Ngiam Tong Dow refused — he commented that it was cheaper to give every student an umbrella.

I called him to say that Mr Lee Kuan Yew had wanted this to be done and that I had to report his rejection. Ngiam told me to go ahead. When I told Mr Lee, he just laughed.

Then, a month later, he went to Darwin. Darwin is a very hot place and they had some sail structures. He sent me a photograph again — I consulted my architect and asked him if we were going to do this. He said no as it was going to be equally costly. Instead, he told me to ask Kenzo Tange to do a simple one.

Kenzo Tange did and it turned out to be one-fifth of the cost of the original lattice form those from Paris. We built it and since it was a design from Kenzo Tange, Mr Lee did not have any comments.

That was the end of the problem.

NTU and the Power of Ideas

Towards the end of the millennium, I planned several ambitious projects for NTU.

I marked $50 million out for a Facility Centre, $10 million for the satellite project, $10 million for the entrepreneurship centre which I thought was necessary to shake our students up, and $10 million to build one of the biggest Wi-Fi campuses in the world. We wanted the whole campus of 200 hectares to be connected.

In 1995, we signed a contract with the University of Surrey to collaborate to send our first satellite to space. We went through quite a lot of hardship because we had people working on the actual setup of satellite in Kazakhstan.

We wanted to build the entrepreneurial spirit in our students and so we ran some courses that were going to be useful for those who want to be entrepreneurs and have their own start-ups. We even had competitions throughout the centre. Through the centre, we have collaborated with some entrepreneurship centres in America. This was a multidisciplinary centre and got students from different faculties together.

Earlier, in 1987, we had already moved to the new part of campus. The former Science Faculty of Nanyang University was renovated and used for the first four years of the engineering course.

The view of the North Spine from my office at NTU, when I was President. I am born in the Yea of the Rabbit, hence the collection of rabbit figurines. My other interest is in transport.

I set up an NTI incubation centre which at any time, we could accommodate 20 external companies. Some were our fresh graduates or staff start-up companies. The only requirement to use the space was to get support from one of the engineering schools. That means they must get a recommendation from the engineering school and show evidence of interaction with the staff and students at the particular schools.

I only allowed them 3 years at one go — if they wished to renew, they would have to show evidence of continued interaction with our academic staff or students.

These companies provided part time employment in the evenings, which was popular with the students. It was mostly engineering student since most of them lived in campus. Later on, there were also accountancy students. It was a good part-time income.

It was a thriving community and some of the companies ended up becoming multi-million-dollar companies. So, a number of professors ended up being millionaires.

The entrepreneurship centre was more for training students. On the other hand, we knew that our engineering professors had a lot of expertise and the government wanted to encourage them to commercialise some of their know-how. In 1998, I was given $1 million by the National Science and Technology Board which was under the Ministry of Trade and Industry to fund staff who had know-how on what could be commercialised. We called it NTU Ventures.

It was true that the failure rate tended to be high. On the other hand, we were careful. I chaired a committee with the director of research and one other senior staff. We vetted proposals from staff that wanted to commercialise their idea and we funded a number which we thought were winning projects.

Within 3 years, we could dish out more money because there were some companies that succeeded and had paid dividends. We would sell out some companies; others were acquired by outside parties.

We took some equity in the companies too. After 3 years, around 2000, we took stock of what we had achieved. We found that we had $10 million to hand out for the engineering department to do research. In 2003, when I retired, I asked the auditors to value the worth of NTU Ventures.

In spite of dishing out $10 million earlier, it was still worth $50 million. Two of the big companies still exist. One is Dense Light, a semi-conductor run by one of the professors who resigned. It is still running today — and had won the EY Entrepreneur of the year.

No outsiders were allowed to invest in the incubators because they were bargains — our professors had a lot of expertise and know-how. So, we commercialised it under NTU and benefitted Singapore. As for the dividends, I gave it to the engineering lab, let them do their own research, build their equipment and infrastructure.

Another big project was the Wi-Fi project. We felt that it was good for such a big campus to be connected. I had visited a university with a small campus overseas. They were quite good in their Wi-Fi connection — I had thought that if they could do it in a small campus, with enough resources, we could do it in a big one.

Out of the $10 million budgeted, we spent around $6 million and saved the rest for maintenance. We were told by our visitors that we were one of the luckiest campus in the world. At 200 hectares, not many campuses so big were fully connected.

Spending this money was for better communication, teaching as well as for the benefit of the students. Now, when people don't have Wi-Fi and network, they feel lost. We already had this in the 90s — so we were ahead of our time.

Business, Medicine and New Frontiers

NTU Executive Centre was built as a facility to host overseas participants including students and professors. When they are in campus, there will be interactions but if they stayed in a hotel, they would all do their own thing after the sessions ended. So, I thought it will be good if they were hosted in the campus.

I visited Harvard and they had a similar facility which they would use to host a two-month business management course. They even had dinner at the dining hall but I thought there was no need for one at our Centre — we could build it next to the recreational centre which already had a dining facility.

First, it was convenient for participants. Second, they would not get distracted by the city. The Centre is now mainly used for short courses, executive program and short stays of academic staff. For short-termers, that was the best place to stay. I had been told that today, it is very popular and not easy to get a room.

Setting up a School of Business in NTU was inevitable for any technology university. It was a strong combination of accountancy, business management and strategy.

Originally when I set up NTU, I saw that at the Massachusetts Institute of Technology, they had strong economic professors in their business school, with a few Nobel Prize winners. I thought we could do the same — so I recruited Lim Chong Yah.

Professor Lim Chong Yah originally a senior professor with NUS who had reached his retirement age. It was useful to have an eminent economist in our business school so when he reached out to NTU, I gladly accepted. He was quite happy because he thought that in the new environment, he could also contribute to the business school. At 88, he remains a professor emeritus at NTU.

Once we had a strong business school in place, we put a strong MBA programme. In fact, I told the Dean that we must make sure that it is unique. Our way of doing this was to work outreach programs into the curriculum. Students had to visit a neighbouring country and conduct in-depth study of a business opportunity and business practice there.

This was not easy because they had to prepare, organise and come back to write a report. To make sure this was done, I joined every trip until after a few years, my timetable did not allow.

In 2000, we decided it was time for NTU to have a medical school by finding a suitable partner. I had a discussion with SGH and we decided to make an overseas trip to see how a technological university can work with a medical school. We went to Israel, then UK's Warwick University, MIT and the University of Washington in the States. These

A mural outlining the key achievements of NTU from the management of NTU to me in 2003 on my stepping down.

3 countries gave us very good examples on how we could convert a techno-university to include a medical school.

As all three places first had a School of Biological Sciences before a medical school, the government suggested for us to start this first. In 2000 I went to New York to recruit Professor James Tam. I interviewed him and found that he was suitable.

So, he came and started our School of Biological Sciences. Being a technology university, we wanted the companies to come in and to provide some real work experience to both staff and students. It was a win-win. We benefitted by working with the companies and they received a stable and qualified workforce.

During my remaining two years at NTU, James built up the school. He did very well and he was particularly interested in TCM so he initiated the TCM course with Beijing University.

Technology does help with medical study. We were already very strong in biomedical engineering — we held many patents and we had people experimenting with guinea pigs. So, we patented the machine which did robotic operations on the pigs. From then, we began to realise it was true that engineering did make a lot of in-roads in medical science and that a partnership with a medical school was good for both sides.

Republic Polytechnic (RP)

I have been actively involved in higher education at the university level for the last 50 years. I have been fortunate in getting myself engaged in the start-up of 3 universities, namely NTU, SMU and UniSIM (SUSS). That provides me with a wide spectrum of the requirements of university learning. In my view, the universities have served well in their roles in equipping our graduates for the present job markets.

Looking ahead, with the rapid technological changes of Industry 4.0 there will be a great impact on our lives, businesses, and industries. We have to re-examine whether the training for our students is still relevant.

From my observations in the last few years, it appears we managed to have a fundamental change in the basis of our higher education. Skill training will be as important as the acquisition of knowledge.

Looking at the types of education in our IHL, Polytechnic training would be getting more important as the time goes on.

Since I have made my own contributions to the universities, I thought it is worthwhile to see what I can do to help our polytechnics. There are five polytechnics with Singapore Polytechnic and Ngee Ann Polytechnics being the oldest. They are well-established and perhaps need little assistance.

Republic Polytechnic is the newest and therefore more open to new ideas. In my assessment, it has the greatest potential to thrive in skill training. It has already done well in working with more than 200 companies to make sure its courses are relevant. I think its PBL (Problem-based learning) is the best approach to skill development and is unique to Republic Polytechnic.

In June 2019, I offered my service to the Principal Yeo Li Pheow. I also asked for his permission to include RP in my book — my email to him is reproduced here. (I thank Dr Tan Lay Pheng, Assistant Director, Republic Polytechnic for assisting me with this work).

****

5 March 2020

Dear Li Pheow

I am now writing my autobiography. As you are aware, I am impressed by your polytechnic's ability to train students to face the challenges of Industry 4.0.

I intend to write a chapter on "Looking Forward" and to quote examples on how some of your initiatives and industrial collaboration centres are doing the right job of preparing your students.

Of course, I can just reproduce some of the information available to the public, but I think it would be better to get Lay Pheng to provide the inputs as she knows the polytechnic's aspiration very well.

I would like to get your permission for her to provide me with inputs.

Thank you.

****

The below are my views on the value of a polytechnic education in the future.

Why do I Recommend a Poly Education?

The world today is changing very quickly, even faster in the last 2 decades with many disruptive innovations, climate change, new infectious diseases with high mortality rate eg. SARS, H1N1, COVID-19. One word to describe today's world — volatile, uncertain, complex and ambiguous (VUCA) and we need to prepare our children for this type of future so that they can have the best chance to keep up with the fast-evolving environment and changes in technology.

They must rise up to take on new challenges.

I look around and scrutinise our Singapore education landscape. In my opinion, a good polytechnic education is one of the answers because it gives you a practice-based learning in a dynamic and progressive learning environment.

In addition to academic knowledge, the Polytechnic emphasises the acquisition of important life skills. One polytechnic in particular has caught my attention — Republic Polytechnic (RP).

Republic Polytechnic is the youngest amongst Singapore's five polytechnics but it has the most progressive and transformative education.

Its unique brand of education, problem-driven and industry-relevant, moulds RP graduates into industry-ready, innovation-ready problem solvers with a strong mastery of knowledge and skills.

RP has a growing track record of grooming effective collaborators, moving in sync with fast-changing world. Through my interaction with RP academic staff, I came to understand RP's rigorous curriculum is aimed at building several important skills — problem-solving, think-on-their feet and time-management. Through collaborating with different team members over the 3 years, students also developed good communication skills, self-management and teamwork.

It is also interesting to note that RP has one of the longest and most comprehensive internships, lasting 5 to 6 months at a company, either local or overseas.

These work internships allow students to gain valuable on-the-job experience, as well as, the opportunities to work with industry experts. From my understanding, some of the students even get job offers after the internship, before their polytechnic graduation.

Of course, many of these polytechnic students do go on to study at universities, like NTU, NUS, SMU, SIT, SUTD and SUSS.

Nanyang Academy of Fine Arts (NAFA)

I got involved in NAFA through an unusual route. Due to a turmoil with its governing Council in 2003, the Council Chairman resigned summarily. Two of the senior staff of NAFA, former MP Ho Kah Leong and Yu Chun Yee approached me to step in as Chairman.

I was not keen as I had just committed to help SIM as its Special Advisor. I was also as an Investment Advisor to the Tan Chin Tuan Group. The latter demanded a lot of my time as there were two morning meetings every week. I declined.

To my surprise, one morning, I was called up to meet Teo Chee Hean, the then Minister of Education. On arrival, I also saw Dr Ng Eng Hen, then Senior Minister of State for Education. They were there to convince me to take up the Chairmanship of NAFA.

Earlier I had already found out that there was deep discord among the Council because of strong disagreements by three groups of Council members. This was typical of institutions associated with the

Chinese community at that time — it had been seen in the Chinese High School and a number of clan associations. Again, I turned down the offer. However, Minister Teo Chee Hean was insistent that I should take up the position, arguing that I was able to deal with the Nantah issue, he was confident that I was able to deal with this one.

He added that if I did not lead NAFA's Council, MOE would leave to stop its funding. I thought this would be drastic and could in fact destroy NAFA. NAFA was and still is the most well-known Academy of the arts both in Singapore and the region. In the end, I accepted the challenge.

I had to make a number of changes to the Council as well as NAFA management.

For a start, all the Council meetings which were previously held in Mandarin are now conducted in English. The two representatives from MOE were very happy as their command in Mandarin was not strong. Another big change was my style of conducting short but effective meetings. In my mind, any meeting which lasted more than one and a half hours is far too long.

Previously, the Council meetings lasted five to six hours starting at 6pm, and often ending at midnight. I was careful that decisions made at the Council meetings were balanced between the rival groups. When I joined the Council there was only one vice-chairman, Poh Choon Ann who ran the Poh Tiong Choon Transport Group.

I decided to balance it by appointing another vice-chairman, Dr KK Phua who ran the World Scientific Group. Knowing that IT and infrastructure were important for an educational institution, I appointed two additional members to the Council, Mr Low Kin Keong, Director ICT, NTU and Hoong Bee Lok, Architect.

There was also a change in the principal from Ho Kah Leong to Choo Thiam Siew who had just left the National Arts Council. One of the main problems of NAFA was that the lecturers were all artists. They were good in art but lacking in governance. Many of the rules laid down were broken consistently by the academic staff.

I felt very strongly for an educational institution to thrive, there must be a strong quality assurance culture. I worked with the Quality Assurance Manager Cecilia Low very hard to get every aspect of NAFA's operation in place with quality assurance.

This implied a lot of hard work by the management and could be difficult to implement for an Art Academy. I am not sure to what extent I succeeded when I left 10 years later in 2013.

Raffles Institution

In contrast, when Minister Tharman invited me to chair the RJC Board in 2005, I did not hesitate. I thought it would be nice to give back what I have gained from RI.

At that time, RJC was run separately from RI. Most of the RJC students in fact came from RI. In 2007, the Integrated Program was implemented which allowed RI students to join RJC without sitting for the O level. This facility was also offered to RGS. It was known as the Raffles Program.

I found it strange that with the Raffles Program in place that RI and RJC remained as separate schools. The fact that RJC received students from RI without ensuring that they were properly prepared was not a good practice. The only way to remedy that was to have both schools run together in a single organisation. I managed to convince both Boards it would be good for the students if the schools could merge.

Mckinsey, the management consultant company, volunteered to do a study, and in the end, they too made the case for merger. I was interested to see if RGS could join the merged school for the same reason but RGS was reluctant to do so, arguing that they were always independent.

In 2009, both RJC and RI were merged into RI. I became the Chairman of RI. It was a pity we could not convince RGS to join the new entity. Now that RGS has moved to Bishan, there is better communication between the two schools. I hope one day, the merger with RGS will take place for the good of the students.

I left the RI Board in 2013 with the satisfaction that I had made a contribution to the outcome of the school.

SMU/UniSIM

Dr Tony Tan left the cabinet in the years 1991 to 1995, but he continued to be advisor to Institutions of Higher Learning, which at the time, was basically just NUS and NTU. In 1995, he came back to the cabinet as a Deputy Prime Minister and Minister of Defence.

Again, he was proactive in driving the university.

Around 1997, he discussed an exciting idea with me: "Look at the US Ivy League universities, all the top universities are private. Maybe we should try to start a university from the private sector rather than being government-funded."

He thought Singapore Institute of Management (SIM) would be a good candidate as it was not part of the establishment. In 1997, he suggested that Tharman Shanmugaratnam and I get into the SIM Governing Council. Tharman at that time was a deputy secretary in the Ministry of Education. But he did not really have a chance to play a big part, because 6 months after he got into the governing council, he joined the Monetary Authority of Singapore.

The SIM chairman was Tan Chok Kian, who had been Singapore's Trade Representative to Taiwan as his last posting. Before that, he had been the permanent secretary of Ministry of Finance. He was into that post for less than 12 months, and by then, was already in his 60s.

Tony Tan thought we needed a much younger chairman for the governing council in order to drive the making of a management university. He called a meeting with the CEO of SIM John Yip, Tan Chok Kian at his office in Bukit Gombak. The director of higher education from MOE (Heng Swee Keat) was also present.

Dr Tony Tan was quick — he outlined the need for renewal and then turned to Tan Chok Kian: "Can you resign as a chairman of SIM?" Chok Kian knew how the land lay and graciously agreed.

At the time, I knew Ho Kwon Ping relatively well. I phoned him to discuss the possibility of him chairing the governing council of SIM. I was very direct, and said: "Perhaps this is an offer one ought not refuse."

He said, "Okay, in a week time, can we meet?" We met at the coffee house at Shangri-La. I told him that I had given his name to Tony Tan who consulted the cabinet, before asking me to make the offer.

He had been surprised by the offer. Then he gave me 2 pages of questions. That meant in one week he had thought it over and prepared quite thoroughly.

I took a look at the questions and said, "Maybe it is not for me to answer your questions. It might be better for you to meet Tony Tan and for you to ask him directly."

I found him a suitable chairman as he was young, in his 40s, and he was in the private sector. He was also quite a dynamic person. To me, the fact that he had once been jailed for his beliefs shows that he was really prepared to speak up — a man of convictions. Edgy, energetic and strong-minded — I thought he would be a good person to start a new university. We elected him as chairman of the governing council.

SIM was very supportive of this idea of a private university because it meant their portfolio would include a university. Most of the governing council members were excited as this was an opportunity to uplift SIM.

The government suggested that we get Wharton, the business school, to help. At the time EDB was part of the government's effort to build links with top universities and institutions across the world.

I went to Wharton together with the current chairman of EDB, Dr Beh Swan Gin, to sign an agreement which allowed some of our staff to be attached to Wharton University.

While there, we had dinner with Janice Bellace who was deputy dean. She later agreed to be the President of the SMU.

I also had the opportunity to decide on the venue of the new campus. Our first choice was the Bayfront area as it was large enough to have all the schools to be together. But URA did not agree, because the land was too valuable. They countered with a site near Outram Road.

Ho Kwon Ping asked, "How about the city?"

To him, being a management university meant being a city university. This is how he said it, but if you had visited the London School of Business you would know that it is a collection of buildings scattered all over the place, so you did not really know where the university was.

But nevertheless, being the chairman, he decided. So, it was developed in the city.

By then I had decided to retire from the Exco because I was running the Nanyang Business School at NTU, so there would be a conflict. It was not an easy decision to leave.

Later, I found out that MOE had decided to take over the university rather than letting it remain under the umbrella of SIM. That is why it changed its name to Singapore Management University.

I asked Tony Tan, "Why?"

He told me that if it remained private, it could not receive government funding. If the government had allowed SMU to get government funding, other private universities would lobby for the same. So, MOE decided to take it over and give SMU an opportunity to develop as a proper university.

I stepped down as a full-time president of NTU at the end of 2002.

In 2003, the chairman of SIM Hsieh Fu Hua felt that SIM's educational arm needed to improve its name and branding and asked me to help. I was appointed as special advisor to SIM Governing Council — a paid position. This meant I had to retire from the governing council.

I felt that the first step was to make sure that education standards in SIM were high. Then it would be respected and be able to attract more good students. So, I started a quality assurance unit.

I spent the next two years raising the standards of teaching at SIM. We came out with a manual, called the "Good Practice Manual" — which distilled my experience at NUS and NTU. The manual I believe, has served SIM until today.

It included practices on the selection of academic staff, conduct of courses, optimal size of classes and also teaching aids. We also looked at the students' qualifications. Over the years, SIM became a preferred institution. Some of the students coming to SIM had rejected offers from NUS and NTU.

Our partners included the University of London and the London School of Economics. The number of first-class honours went exponentially up year after year — that is an indication of how successful our education is. Today, we have about 200 first class honours. They take the same exam as LSE and in fact, I met the LSE representative who told me that we have more first-class honours than LSE itself.

I did a sampling of some of the first-class honours and found that from their A-level results, they could have gotten them into NTU, NUS. I had a meeting with about 10 of them and asked "Did you apply to NUS?"

All of them said, "Yes, but we rejected them and came to SIM."

Why? First, it was cheaper, second it is a degree from a highly-ranked university. Most sensible reasons I thought.

In 2005, Tony Tan spoke to me.

He said, "Up till now, we had not really been serious about lifelong learning. We do not give second chances to working adults who have not had the chance to study for a degree. Government has just paid lip service."

He encouraged me to start a university to provide part-time learning for working adults. That is how UniSIM was born.

We placed emphasis on courses that would be useful in the market. For example, we had joint courses with the police, with MINDEF, which cater to their requirements.

We were also very big on e-learning. Working adults find it hard to come to evening courses, especially as a lot of workers worked late or had to travel overseas for work. E-learning enabled them to learn anywhere and at their own pace.

At one of our graduation ceremonies, I had the chance to talk to one of our graduates. He said, "I am very lucky and happy that I graduated from the midnight university."

I asked, "What do you mean?"

"I worked in daytime, when I got home, the family came first. When the children and the wife were asleep, I started my class — that was already midnight. And I could do it because we had a lot of e-learning."

Not long ago, Ministry of Education looked at the Institutions of Higher Learning and ranked UniSIM as the most advanced in e-learning in Singapore.

It was not easy to impose e-learning in the university. The professors — whether part time or full time — were brought up using markers and whiteboards, face-to-face classrooms — e-learning is alien to them.

Somehow we had to get them to accept their idea that e-learning is the pathway for the future.

Celebrating the second year of SIM University

It was difficult — not so much because of hardware which in any case was not expensive. It was actually the software and training the teachers. We had 600 to 700 part-time lecturers — they had to pass an assessment to able to conduct e-learning before we allowed them to teach in class.

It was hard work, but over time, we succeeded in making it the second nature at UniSIM. We are now quite well known all over the world for our e-learning program.

Of course, there is no UniSIM now — the MOE

took it over and renamed it SUSS — Singapore University of Social Sciences.

The government wanted to focus on lifelong learning, especially those who wanted a second chance at getting a degree. The MOE nearly wanted to start their very own open university. They came to the conclusion, that if they did start one, they could not compete with UniSIM. So, it was easier to take it over.

The SIM Governing Council, of course, was reluctant to give it up. I said, "I don't think we have any way to resist, because we could not award degrees without MOE's agreement." So, my advice for the governing council was to accept their decision.

This was the second time we lost a university as far as SIM was concerned. After 9 years of being involved in SIM, I retired.

There is an interesting story which I must add.

In the beginning when we became UniSIM, a full-fledged university, we needed a chancellor, I approached SR Nathan and he agreed. This was in 2005, when he was President of Singapore.

One day, I got the call from Tharman, he was then the minister of education.

He told me, "You want SR Nathan to be the chancellor. MOE cannot agree to that." I asked, "Why?" He explained: "You are a private university. Other private universities had asked him to be the chancellor. We had not agreed."

So, I said, "Yes, I can see your point. So, who ought to be the chancellor?"

He said: "You."

So I became both chairman and chancellor. It is very unusual, but the Minister of Education had suggested it.

UniSIM was a private institution so the Government could not support us directly. What they did was support the students. So, the students who are Singaporeans, for example, they get 55% of the fee subsidized by government.

Our revenue comes entirely from the fee of the students. There were no government injections. But of course, UniSIM being part of SIM, we had some help from SIM itself. Because SIM, having started in 1964, had accumulated a lot of funds.

So, over the years, they gave money to UniSIM, because they are proud that UniSIM is a university under the SIM umbrella. For example, bringing in new equipment, software, getting people in to develop some of the e-courses — these are additional expenditure coming from the generosity of SIM.

For operating expenditure, we depended on the fees of the students. And, of course, they in turn get subsidies from government.

We have a small number of full-time staff, but a lot of part timers. They come from industries and businesses, and are very knowledgeable. The only problem is that sometimes their teaching skills are not quite up to par. So, they have to attend teaching seminars and training. We also get the feedback from students at the end of each course. If we find that the lecturer is consistently below par, we will not continue to employ them.

I would say 50% of our full-time staff, have a PhD. The rest have at least a Masters degree but all have strong experience. In fact, even the PhDs have got good industry experience.

It is not difficult to recruit staff, especially part timers. Many have actually been working 10–20 years some of them own their business and are surprisingly altruistic. Those associate lecturers want to share their knowledge with younger students.

In fact, I think it is more difficult to attract good lecturers to work full-time because we need people who are good in teaching and good in research.

In general, students took about 5 years to complete a degree — a bit longer than the normal 4 years, because they are working and find it hard to take the required modules. The drop-out rate is also a bit higher, about 20, 30%. Again, if their work is too heavy, they will not be able to complete.

In fact, I think compared to the many overseas part time degree, 20 to 30% drop-out rate is considered good.

In Singapore, everybody is hungry for a degree.

Our target is to have about 3,000–4,000 graduates every year, which is a lot since our student body is around 15,000. Our range of ages is very wide — the oldest being 65. These older students are here more for personal satisfaction than practical reasons.

Quite a large percentage get a jump in either position or salary after getting a degree. We also get a few disgruntled students who say that, "My employer does not recognise my degree." But, in general, many do benefit from the education.

Employers' perception of UniSIM degree is very good. Our students are very savvy. There are some full-time students who tend to aim for first-class honours. But most people who are much more realistic and the working students already have some insight into their jobs and are more confident.

They do not choose easy subjects. The full-time students are the ones who tend to choose "easy subjects" to get a good shot at first-class honours, but working students choose subjects that they think are relevant to them.

They are very dedicated. I take off my hat to them — study, work and family, how do they manage all these? I think their success is doubly valuable when compared to full-time students.

I was brought in to launch UniSIM. But SIM also decided to nurture entrepreneurship in young students. We think that, very soon, most of our graduates would be either self-employed, or working for small companies.

So, they need to more of an entrepreneurship spirit. So, I have been asked to help a new centre — platform-E — the E stands for entrepreneurship.

Incubators are very competitive in Singapore so why should someone choose SIM? They will come if we have got the funding. I went to Silicon Valley to try to attract some of the venture capitalists, angels to come to SIM.

Many young people need a lot of knowledge about how to work in small company. For example, they must learn to be able to look at the balance sheet and finances, how to do marketing, to employ people, what laws are involved. They must know how to use IT.

These are some of the basic skills that fresh graduates are ignorant about. If you are an art or science or even engineering graduate, you know your own discipline, you know nothing about these practices. When you go to a small firm, say 3 to 4 employees only, you will have to do all these. This part of education was also very important and useful.

Every year in Singapore, we have 15,000 graduates. Maybe 2,000, 3,000 work for the government, and big companies. But the other 12,000–13,000 have got to fend for themselves. So, they must know all these skills — we tend to emphasize technical skills.

To me, the more urgent demographic is our fresh graduates — how they can take an idea, run with it and make a good living out of it.

Purcell School, London

Purcell School in London was started in 1962. It is the oldest school for gifted children in music in Britain, for students between 9 to 18.

The enrolment has always been kept small at about 200 students. Other than music studies which is compulsory, students have to take a number of GCSE subjects. Over the years, it has attracted many Asian students particularly from South Korea.

My daughter Gee Len was enrolled in the school from 1992 to 1994 at the age of 16 after passing through an audition. She is a pianist and had already passed the highest grade of LRSM of the Royal School of Music. She lived with a family whose house was near the school.

I visited the school a number of times when my daughter was a student. It was at such visits that I got to know the Headmaster. In 1995, he invited me to be a Governor of the School. His reason was there were about 25% Asian students, and none of the Governors was Asian.

In fact, all of them were British and were prominent musicians. At that time, I was the President of NTU and also an active member of our SSO Board. So, it was appropriate for me to accept the position. It would also allow me to be exposed to the working of a foreign board and in an area of great interest to me.

I retired five years later in 2000. On reflection, it was an exciting experience for me and I am grateful for the opportunity.

It was a well-known school producing many prominent musicians such as Robert Cohen (cellist), Oliver Knussen (composer and conductor) and Yevgeny Sudbin (pianist). In fact, Sudbin performed in Singapore in January 2020.

Chapter 4

Doing Good (Societies & NGOs)

Rhodes Scholarship

The Rhodes Scholarship was established in 1902 and is the oldest and perhaps the most prestigious international scholarship program in the world. The scholarships evaluate outstanding people from around the world to undertake postgraduate study at the University of Oxford.

One of the founding aims as stated by Cecil John Rhodes "was to identify young leaders from around the world who, through the pursuit of education together at Oxford, would forge bonds of mutual understanding and fellowship for the better of mankind."

Each country has its own selection committee consisting of five persons. It has an unusual requirement that the Chairman cannot be a Rhodes Scholar but the other 4 have to be former Scholars. In 2001, I became the Chairman by virtue of being non-Scholar. We used the following criteria for the selection of Scholars:

(1) Scholastic attainments. We did not demand the candidate to be top of the class but had to be 'A' students. Part of the reason was that the scholar has to subsequently undertake postgraduate studies at Oxford and therefore need the basic scholastic abilities. The scholar can also then devote some time for bonding with fellow classmates.
(2) The candidate has to excel in certain talents as demonstrated by mastery in specific areas such as sports, music, dance or visual art.

(3) Leadership quality as demonstrated by ability to lead a team, contribution to national projects or being the head prefect of a school for example.
(4) Moral courage as seen by the candidate's interest in fellow beings or charitable organisations.

While some criteria are easy to assess, others required judgement and intuition. In the 6 years I was Chairman, our decisions had always been unanimous.

Notable Rhode Scholarships for Singapore are Dr Tan Eng Liang, Mr Raymond Lim (Ex-Minister) and Chen Show Mao (MP, Workers Party).

In 2006, Singapore ceased to give Rhodes Scholarship, but was reinstated in 2019.

The Head Foundation

The Head Foundation (THF) is a low-key foundation set up by a wealthy Indonesian, Sendi Bingei. It aspires to be a thought leader for key issues that affect Asia and its people. It supports the sustainable development of Asia through education and healthcare.

THF was registered in Singapore in 2013 as an International Charitable Organisation. It is supported through the philanthropic donations of its sponsors. Working with a network of public institutions, the Foundation sponsors policy, pedagogical and clinical research to help educate the development of education and healthcare in Asia. Research findings are published and shared with government authorities, practitioners and the public in the relevant countries.

To disseminate knowledge and share ideas, it hosts regular public lectures during which new ideas on education, environment, science, culture and healthcare are presented and discussed. These lectures are recorded and shared through YouTube for viewing. This series of lectures are well received by the public in Singapore.

In 2018, at the suggestion of the sponsor, the Foundation received its mission to explore larger scale projects that would be sustainable, replicable, and measurable.

To assist in charting new directions and initiatives, an Advisory Board was formed to provide advice on the selection, funding and implementation of large projects which would include research into East–West integrative medical treatments. I was appointed to Chair the Advisory Board. Two other members are Prof Robert Pho, an eminent orthopaedic surgeon in Singapore and Dr Wolfgang Schuster, an ex-mayor of Stuttgart, Germany.

Additional members would be added in future. We are to meet twice a year in Singapore.

Since the formation of the Advisory Board, we have met twice, once informally in a retreat conducted in Wuyi San in January 2019 and in October 2019 in Singapore. The task of the Advisory Board is to advise the Board and management on any issue raised by them and any new direction THF should take. The Advisory Board also helps to review larger projects of more than half a million dollars.

Over the past years, many well-known academics all over the world have got in touch with THF and shared their views and experiences. They were considered as Fellows of the Foundation. Recently, the Board decided to formalise the appointments by having a smaller number and a good spread of countries. A small honorarium will also be given.

I am glad to make my contributions to THF as it is willing to do good for the society especially for the region. It can be a catalyst to inject good policies to society and country. An example is the recent interest in environmental sustainability.

Singapore Symphony Orchestra

I have a great interest in music. In my primary school, there were a few recorder ensembles and I was good enough to be a soloist in recorder playing. By the time I went to secondary school, I was much more open to music of all kinds.

Both of my sisters were given the chance of having a piano teacher. But my mother refused to let me have lessons. She said boys should not waste time on music! However, I was very interested whenever I had the chance to hear my two sisters play. Since there was a piano at home, I played with it whenever they were not using it.

I began playing while I was in Secondary 2 and had started attending concerts. In the past, SSO was called the Singapore Music Society. The music director, as well as conductor, had come from London.

One of the pieces that really caught my attention was Symphony No 8 of Beethoven. When I was in Secondary 2, I was introduced to music and later got to know Symphony No 5.

The concerts were in the Victoria Memorial Hall. There was no air-conditioner at all back then. The cheapest seat was at the back, just right behind the stage. Being a student, I did not have much pocket money, so I ended up sitting at the back.

Now, we call it the "quiet seat" — while back then we called it, "can't remember where we sat seat". They were the cheapest seats, but for me at least, they were the most treasured because among these cheap seats was where I met my wife, who had attended a concert with my classmate's sister.

She loved music as I did and so we had many opportunities to meet. We had to buy the tickets in advance. As we were at the back rather than the front of the stage, we could not get the best views — but we were very close to the performers.

One memorable performance I really appreciated was Marian Anderson. Marian Anderson was actually a well-known contralto. She sang to the audiences to the front of the stage. However, she knew that there was a group of listeners at the back so during one of the encores, she turned around and sang towards us. I was impressed, because when she turned around, I could feel the vibration as she sung the vibrato. That is how powerful she was. She died in 1993, but her performance that day stayed with me for life.

At that time, a renowned pianist appeared on the musical scene — Claudio Arrau. Arrau played on that stage as well as another pianist, Eugene Istomin.

When all these big stars came, the price of the tickets were still $1. I was amazed at how they did it because they were all in jackets, and there was no air-conditioning. They all performed for one whole night! Now, even with an air-conditioning on, I can still see some performers sweat in their jacket.

Members of the Musical Society were mostly foreigners. Most students who attended concerts were also Caucasians. The concerts were well-attended because during the 50s, if you were an expatriate, there was not much in the way of cultural exploits in Singapore.

I don't know how they manage to get all the top-class musicians to come. In fact, I still remember that there were occasions where I sat next to Yu Chun Yee, a pianist. He is a well-known pianist and complimented the performer on stage for his amazing technique which captivated us both.

Thinking back, these concerts were among the greatest pleasures of my lifetime. I bought the cheap tickets but we still had the chance to get close to the performer. They were all the top artists, even violinist Ruggiero Ricci. That was how I fell in love with music — watching the best perform.

Not surprisingly, I wanted to be a musician, but my father put a firm stop to it. He wanted me to be an engineer since I was good at Mathematics. So that was the end of that conversation. But I never stopped listening to music.

At the time, Lee Hsien Loong was the deputy prime minister, he came to know about my love and knowledge of music. In 1997, the government got me on the board of SSO, together with conductor Lan Shui. The year after, I was made deputy chairman. The Chairman at that time was Tan Boon Teik. I went to board meetings but very often, there would be no decision at the end of the meeting.

The Guest of Honour at SSO's 30th Anniversary was Prime Minister Lee Hsien Loong.

It is not easy dealing with musicians and artistic issues. When I joined the board, everybody seemed to be doing their own thing. When I became chairman a few years later, I did my best to make sure the board itself was organised.

The departure of long-time Maestro Choo Hoey was a difficult decision.

From what I understood, some musicians did not get along with him. One musician complained that were occasions which his instructions were confusing.

While he had good knowledge of music and conducting, some felt he was not up to the task of leading the orchestra.

When we started SSO, the government wanted a Singaporean conductor — he was the best that we could get at that time. Even by 1997, I would say that, if you want a Singaporean, he was still considered by some to be the best. But if we truly wanted SSO to make a leap ahead, we needed someone superlative. For that — I thought of Lan Shui.

Lan Shui joined the SSO in 1997 and brought in a new era of structure and discipline.

He came in at the cusp of a transformation. Our attendance was growing. The management too was changing.

I made sure that our meetings were constructive. We had a list, went ahead and decided on what needed to be done and then got the management to implement. I am very task-oriented. I was reminded of my time at NAFA, where we went through the same thing.

Gradually, we became more structured and disciplined.

In 2004, I got Chng Kai Jin in as SSO's general manager. At the time, we were in a deficit of about $6 million — slowly, we brought it down. Today, at the Esplanade Concert Hall, our concerts are about 85% full on average. Kai Jin's strength was that he was very people-oriented and had good EQ.

My intention was to retire after 9 years. By 2008, I knew I wanted Goh Yew Lin to succeed me. However, Yew Lin said that he was busy during that period as his stockbroking firm GK Goh was undergoing some restructuring.

He asked for another two years so I agreed. But I soon reminded him that I needed to step down. He wanted to retain me so he created the SSO Council and asked me to chair the Council.

The Council had a strategic mission to look at what we should do and how to get new sources of funding. I was glad that I was able to get eminent people in my council — JY Pillay, former SPH CEO, Alan Chan, and Stephen Riady. They were all supportive of the arts.

The home of the SSO is now at the Victoria Concert Hall (VCH). We had all assumed that when Esplanade was built, it was to be the home for SSO. However, the Minister at the time, George Yeo, made the deliberate decision that Esplanade will *not* be the home of SSO.

It was a great disappointment to us because putting out a concert, we need a few days for rehearsals, so depending on how the progress goes, we may need more days. But sometimes we find that it was not so easy because the Esplanade already had other customers and bookings come in.

There was a great sense of frustration because either we did not have the date that we wanted, or we could not meet the programming schedules of artists coming from abroad. They normally do not give you too many options, and we needed the flexibility which, perhaps understandably, the Esplanade was not able to give.

So, while it was not quite what we wanted, the SSO needed a home, so we took the Victoria Concert Hall since we actually owned it. The trouble is SSO decided on this only after the renovation — so we did not have as much input into design matters and acoustics as we did for the Esplanade.

Over the years, the VCH which is a beautiful venue, has given us some good income but it is not an ideal place for large symphonic concerts.

The SSO brought in world-class musicians, including Lang Lang (pictured here with his mother) and Chairman of SSO Goh Yew Lin (far left).

Chapter 5

Engineering Institutions

Institution of Engineers, Singapore

It was because of my father's encouragement that I became an engineer. At that time when I made the decision, I knew very little about engineering. Looking back, I am glad I made that decision.

Engineering teaches you how to apply basic principles to solve problems. That is applicable to every aspect of daily lives. In life, there is never an ideal solution. It is always a compromise, very similar to solving engineering problems which have no exact solutions. For me, having been trained in the exacting fields of Mathematics and Physics in school, it took me a while to get used to that philosophy of imperfection in my early years in the University.

I did well in my first degree in the University of Malaya. The course was quite general, offering me knowledge of almost every aspect of engineering practise. By that time, I decided that I wanted to be a research engineer. Therefore, I got myself a place at Cambridge to do a PhD in the high-powered field of aerodynamics.

On return to Singapore in 1968, I was disappointed that there was no research culture at that time. It would be futile for me to embark in a research engineer career. The only course of opening for me was to teach engineering at the University and practise engineering work by giving consultancy on specific problems. I thought it would be wise for me to join the Institution of Engineers, Singapore (IES) as a member so that I could meet fellow practising engineers.

IES is the national society of engineers for the advancement of engineering in Singapore and was established in 1966 as a breakaway from the Institution of Engineers, Malaya (Johor Bahru branch) after independence. IES, being the premier engineering institution is often called upon by the government to provide feedback on professional engineering matters. As a consequence, IES members sit on a large number of government committees. It works closely with the IHLs to provide useful industrial input to their engineering education.

Initially, I had difficulty being accepted by IES as a corporate member. According to its constitution, corporate members have to acquire at least 3 years of working experience in an engineering enterprise acceptable to IES. Teaching at a Faculty of Engineering was not accepted as equivalent to practising experience. This was not so at a British Institution of Engineering. At one of the meetings of the Faculty of Engineering, University of Singapore in 1970 with the then Vice Chancellor Dr Toh Chin Chye, he discovered that I was rejected as a member by IES. He called for a meeting with the Council of IES the following week and told the Council he was unhappy with IES not accepting teaching experience of lecturers at the university as equivalent to practising experience.

He thought IES could make use of the talents of lecturers who were normally very qualified, many of them PhDs. As a result of the meeting, IES offered me a Corporate Membership.

Having been admitted as a member, I decided to play an active role in IES. I stood for election for a position as a Council member in the Mechanical Engineering category. My votes tied with Mr Tay Sin Yan, an existing Council member. He was kind enough to give way to me and so I became a member of Council from 1971. Over the next

15 years, I remained in the Council, initially as Ordinary member, then Hon Treasurer, Hon Secretary, Vice President and as President from 1980 to 1982.

Over this period, I saw the membership increase rapidly initially from about 1,000 to 3,000 by the early 1980s. Although the majority of the members were from Civil Engineering, we expanded to Chemical Engineering and Marine Engineering.

The Secretariat of IES was housed in a 400 square feet office in 1971 at the Professional Centre in Outram Park. It became very apparent the space was not adequate and so in 1973, IES purchased a new office of 221 square meters in the International Plaza in Anson Road.

Despite the larger area, the space and facilities were insufficient for the growing membership. In 1983, IES moved to a purpose-built house on Bukit Tinggi Road next to the Swiss Club with donations from members and the business community for the 6,800 square meters plot of land.

During my time as an active Council member, other than providing facilities to our members, we began an active role in cross-border relationships. Ties between IES and its Malaysia counterpart continued to develop. The two institutions decided to promote interaction among their members through a joint IES/IEM Engineering Convention. The first was held in November 1974 in Singapore and the second the following year in Kuala Lumpur. The Engineering Convention was the creation of both Councils which met every six-monthly since 1973.

Following the success of the two conventions, we decided at the third Engineering Convention in 1977 to invite engineering institutions in other ASEAN nations to attend.

At this convention, the heads of delegations agreed to establish a Council of Engineering Institutions of South-East Asian Nation

(CEISEAN) and a convention would be held yearly in various regional capitals in turn. It was at the second CEISEAN meeting in Manila in 1980 that the ASEAN Federation of Engineering Organisations (AFEO) was formed to replace CEISEAN. This process of establishing contact, good will, rapport and cooperation amongst engineering institutions was a good development for engineers in the region. While IES is also a member of the Commonwealth Engineering Council, it was not that active because of the great diversity of members which made it difficult to identify common ground.

I gained a lot in my personal development while being an active member in IES. The Institution being a professional body gave me the exposure to all types of professional engineers. The knowledge gained was useful in my future roles in the boards of companies and other bodies.

After being the President of IES in 1980 to 1982 gave me the confidence to run NTI/NTU, especially since the initial schools were in engineering.

Professional Engineers Board (PEB)

The Professional Engineers Board was formed in 1971 through an Act in Parliament. It was tabled by Dr Toh Chin Chye then the Minister for Science and Technology. This was resulted through an appeal by IES to Dr Toh to register professional engineers.

PEB certified professional engineers as being qualified and competent to design and plan complex structures and machinery, supervising the construction of those structures and maintaining the safe operation of machinery. The Professional Engineer is accountable legally to ensure public safety. With the Act, we had teeth.

A major provision in setting up the PEB is the task of registering professional engineers. In the early years, since there were no ready-made criteria to be used, the Corporate membership of IES automatically qualified for registration.

In 1977 this was discontinued, as a proper Qualification Committee was formed to decide on the acceptable degree training and working experience. For some cases, candidates had to pass a practice examination.

Degrees from NUS and NTU were accepted without the need of further examination. It was also arranged that 3 nominees from IES would be appointed as Board members of PEB by the Minister for National Development. This move signified the close association of IES with PEB.

I was the nominee of IES and served on the Board of PEB for 6 years from 1973. I also served as the Chairman of the Qualification Committee.

The setting up of the PEB is good for the engineering community. It recognises the contribution of engineers to the development of Singapore. At the same time, it ensures a high standard of engineers and gives us pride as well as a sense of community.

Academy of Engineering, Singapore

Many countries including those in ASEAN have an Academy of Engineering in one form or the other. Singapore did not have one until 2011 when a member of environment engineers decided we should form one to provide leadership among engineering professionals, to advance engineering education and to promote engineering excellence in general.

The Academy has elected 114 prominent engineers to its Fellowship. Fellows are elected through a vigorous peer-review process which evaluates individuals on the basis of their leadership and career achievements. About half of them are in the private sector heading many top companies and organisations.

The list of Fellows reads like a who's-who among engineers in Singapore. They include eminent engineers like JY Pillay, Lee Ek Tieng, Kwa Chong Seng, Chew Choon Seng, Wong Nik Liong, Koh Boon Hui, Moses Lee, Ho Ching, and Lee Seng Lip.

There are two roles which the Academy can play in Singapore.

First, the Academy can identify strategic issues and projects and call upon its pool of expert Fellows to deliberate and lend their expertise. Their deliberations can end up in a consultation paper which can be published in our web page or newsletter. It can also be sent to interested parties for their information and reviews. Three such papers are given below:

(i) Singapore as a connected city. Cities are growing at a rapid pace. The increase in population density and human expectations of better economic and social lives have made the planning and governance of cities of the future an issue of extreme importance. Advancements in technology can facilitate the creation of connected cities. Such cities allow easy access to data and information, make extensive use of computing and information technologies in the operation of its physical infrastructures as well as social services, and in its economy and industries. A roadmap for the future of Singapore as a connected city is also presented. (Refer to website for paper)
https://www.saeng.sg/key-projects/singapore-as-a-connected-city/

(ii) Improving sustainability in Transport Systems for Singapore. (Refer to website for paper. Led by Prof Lee Der-Horng)
https://www.saeng.sg/key-projects/improving-sustainability-in-transport-systems-for-singapore/

(iii) Renewable Energy Alternatives for Singapore. (Refer to website for paper. Led by Prof Chou Siaw Kiang)
https://www.saeng.sg/key-projects/renewable-energy-alternatives-for-singapore/

Second, Fellows can serve as role models for the young to show that they can excel in many fields eventually. We hope that would encourage more bright students to study engineering.

I have been an engineer all my life, and received the International Medal from the Royal Academy in the UK from Prince Philip (2006).

A closer look at the International Medal and plaque.

Chapter 6

Government Bodies

Science Council

My academic life was a series of fortuitous events. I entered engineering not knowing anything about it — in fact, I only knew about engineering during my first year of university studies.

I started in civil engineering because at that time that was the only course for engineering at the University of Malaya in Kuala Lumpur. I did not find civil engineering very challenging as it involved a lot of common-sense work such as how to construct good buildings and how to lay good roads. After finishing civil engineering, I got admitted to Cambridge and decided to go into aeronautics which was different and challenging. My thesis was on spinning missiles, a fascinating and exact part of engineering.

My professors asked me what I would do after I returned, because Singapore did not have a missile industry. I decided to join the university.

The Science Council was started by the Ministry of Science and Technology. In 1968, the Minister was Dr Toh Chin Chye. For some reason, I got on very well with Dr Toh. In 1972, he wanted me to start an ocean engineering centre for Singapore.

Dr Toh had thought that studying ocean engineering was something important for Singapore as a port and an island, and would stand the country in good stead far into the future.

Today, we are talking about drones and missiles but those days, we were talking about the oceans which Singapore could explore and exploit. Therefore, he wanted me to start the ocean engineering centre.

He sent me to Japan, Nagasaki, the heart of the Mitsubishi enterprise. Mitsubishi had already a shipyard in Singapore and that they made use of the former British Naval Base. In 1972, I went to Hawaii to see what they did and again, they also had some ocean engineering efforts.

Dr Toh put me into the Science Council in 1975. In 1981, the Ministry of Science and Technology closed as Dr Toh left Government, although he remained a member of Parliament. From 1981 to 1991, the Council was under the Ministry of Education.

The one who really looked after the Science Council and initiated a lot of programmes was the Chairman, Professor Ang How Ghee, together with Prof Choo Seok Cheow, who eventually became one of the Deputy Vice-Chancellors of NUS. Dr Aline Wong a sociologist was also a member of the Council, before she became Health Minister.

That was the first time I met her and we remained very good friends up to today. Of course, Professor Bernard Tan was there. These were the four that I remembered.

Our main mission was to get people to be interested in science. There was a science competition among schools on television "The Science Quiz with Professor Bernard Tan". It was an initiative to interest students through interschool science quizzes.

They also invited very senior scientists to give talks in Singapore — one of them was Sydney Brenner, a Nobel Prize winner who went on to have a long and fruitful relationship with Singapore. One thing that the Science Council did outstandingly well was forming the Science Centre. It was very much welcomed by the schools, schoolchildren as well as parents. That included many eminent scientists coming over and through the cooperation of the Royal Society of Britain.

In 1991, the National Science and Technology Board was formed under Ministry of Trade and Industry. The Science Council was dissolved, because we realised that even in the early 90s there was really no science policy in Singapore. The science policy was a derivative of an economic policy.

NSTB's role was to bring high tech companies and technology to Singapore for economic development, rather than developing scientific knowledge. The mission to inculcate love of science was taken over by the Science centre — through hands on activities and demonstrations. This is the case now in many countries all over the world.

NSTB

As I was then the President of NTU, we tried to map out some of the work that would be useful to support the economy of Singapore. We made sure that our graduates were well trained in those areas. I also made sure that our research was relevant — NSTB being part of MTI had more funding, than the Science Council under MOE.

Consequently, we could embark on many more research programmes.

Among the most important were electronics and wafer fabrication. Singapore was becoming to be a wafer fab centre in Southeast Asia and maybe in the world. We attracted many big wafer fab companies. Also, we foresaw that electronic export would be a very important part of Singapore. A third area was automation. It was simple in the beginning, using computers to facilitate automation, very simple robotics to do fabrication to make our electronics industry competitive.

Funding was generous. However, the problem was getting enough people interested. There were very few Singaporeans in research, so there was no critical mass. They were more likely to go into areas such as finance, banking, law and medicine where salaries were higher. Hence, we found that the only way for people to do research was to get foreign talent.

We recruited from all over the world, including from China, India and Malaysia. People would come in to do PhDs as part of the research programme. We managed to attract some Singaporean undergraduates to consider research or academia as a career. Some do come into research but very few — that's the same, even now.

PhD students were mostly foreigners — only 20% Singaporeans.

Going into research is a bit like a religion, a firm commitment, not something which you go into thinking you are going to be famous and make a lot of money. Those who succeed in becoming rich are just very lucky. Only one in a trillion will succeed like Steve Jobs. On the whole, scientists and engineers are not going to be rich.

The National Science Awards was part of trying to enthuse the young, especially Singaporeans. It was NSTB board idea to start the awards. At the beginning, the awards were confined to just Singaporeans and PRs. Later on, it was expanded to include all nationalities as long as there was a Singaporean in the team.

But some rewards are beyond price or prestige — for me, the science itself is the reward. However, when I came back from Cambridge in the 60s, there were hardly any research. I ended up doing very practical engineering and at the end I was being roped into being an education administrator.

There are now 6 universities in Singapore, out of which three had been started by me. NTU, SMU and then of course UniSIM, now called SUSS. Therefore, I ended up developing my career as an education administrator, or educational midwife.

So, I never really got to do research. Perhaps, it was just poor timing.

When I finished my PhD, I nearly went to the United States because of my work on spinning missiles, my thesis topic. There was a defence company in the States, Lockheed, who recruited me. They had already made all the arrangements so I told my wife that we should pack up and go to the States.

On the point of departure, I was told by the Department of Defence, that I could not clear my visa because in 1968 it was the height of the Vietnam War. My surname was "Cham" — there was a place in North Vietnam called Kampong Cham and they thought that

Finally, many years after I was denied a visa to the US because of my surname, I visited Kampong Cham, Vietnam (2002).

I was related to the North Vietnamese. They refused to clear me — so I came back to Singapore.

When first I came back, our Ministry of Defence was most interested in roping me in. At that time, the Chief of Defence was Kirpa Ram Vij and he was the first one to interview me. The MINDEF headquarters was in Pearl's Hill at that time.

Ministry of Defence had started the first defence research agency called Defence Science Organisation. But by that time, I decided that I wanted to have an academic career.

As I said, although I loved science, I never did as much as research as I wanted. But I have no regrets — I think my life has been richer for the different path I took.

Nowadays, missiles no longer spin because they are guided by laser. In the past, they spun to keep flying straight. In a rifle, the muzzle is not straight-barrelled but has a groove inside so the bullet comes out from the rifle spinning, flying straight through to the target. That was the reason why in my time, a crucial element of a missile was its spin.

Sometimes, in life too, you need a bit of spin to get to your target.

A*STAR

A*STAR came in 2000, a few years before I retired in 2003 as the President of NTU (but I have been called into service in many places since then — so you could say I retired many times).

I was a member of Biomedical Research Council in 2000 and the reason I was roped into A*STAR was because we wanted to make sure that NTU was able to support its work. The Chairman was Philip Yeo and then Lim Chuan Poh.

A*STAR's job was to make our industries more competitive — it left the universities to do research and it instead created research institutes. Each research institute specialised in certain things. I started one in the late 80s — it used to be called GINTIC. Now, it's called SIMTech. GINTIC stood for Grumman International-NTI Centre.

GINTIC researched how computers could be used to help in manufacturing so that it was more precise and automated. It was called CAD-CAM, Computer Aided Design-Computer Aided Manufacturing. In the late 80s, we identified that as an area that would be good for NTI.

Grumman Aerospace was quite advanced in this area — so I went to New York and got them in as a partner. We knew very little about the subject and had no practical experience. I negotiated and signed a contract to service for five years.

Grumman Aerospace gave us both hardware and software and they got two or three persons to come and train our people. In the meantime, I also recruited our own staff, including a Dutchman — Frans Carpay. He was working for Philips at that time and I poached him to come to Singapore.

There was a cultural difference but he was quite good in getting it running in NTI. Later, A*STAR took over and they made it into an institute which they funded — part of it is still in the NTU campus.

GINTIC started originally as basically a training institute and we worked with industry from the start. I recruited Dr Carpay for that purpose and he even went with me to China to recruit researchers .

Science and technology are fascinating topics to me. Researchers and scientists always seem to be a little bit ahead of our time.

In the 1990s, in NTU we had a 3-D printing machine. We called it a "prototyping machine". It read from the computer then "printed" a solid object using resin which was a plastic material. So, we already started 3-D printing in the 90s.

It was not an experiment — we used it for teaching and doing research. We produced souvenirs for our visitors at that time. We were ahead of our time. Even now, it is not easy to have a commercial application. First, it was very slow and very time-consuming to prepare. Secondly, it was very expensive. So, it was used only for very specialised products.

As an industry, I do not think it will come about so easily — maybe give it another 20 years.

A second product is the autonomous car, something that I am not very sure I can see in my lifetime. A real autonomous car would mean that I can step into the car and say "Bring me to Bukit Timah Plaza" and the car would drive me there. Again, that will come maybe in 20, 30 years' time. It is still science fiction at this stage.

Third, financial technology. It started quite long ago from the time of ATMs. I have been on the board of UOB since 2001 but UOB had ATMs since the 1960s. When we introduced ATM in a big way, we thought we would recruit far fewer tellers. Even today, we are still recruiting tellers although fewer because there are just some things you need to talk a person at the bank about. You cannot just talk to machines.

FinTech is a buzz word. Everybody talks about it like it only happened recently but it had been happening for a long time. It takes time, just like 3-D printing which already started in the 90s.

We now think that artificial intelligence can replace people. It is not so easy, it takes time to evolve and by that time people would have changed job already if they were nimble.

AI is not as disruptive in the sense of bringing about overnight changes but in the sense that things would be different. It would take place slowly and there would be time for people to adapt and to evolve and maybe learn new things.

Construction Industry Development Board (CIDB)

In 1984, I was chosen to be Chairman of CIDB. Perhaps it was due to the fact that I had been President of the Institution of Engineers and at that time was running a successful NTI.

CIDB was a statutory Board of the Ministry of National Development. Members of the Board consisted of engineers, architects, contractors and developers; there was also an auditor from PWC.

The role of CIDB was to promote the construction industry in Singapore with special emphasis on construction quality and productivity. During my tenure, there was an economic downturn in Singapore — so an additional role of CIDB was to promote construction business in the region, especially in China and Taiwan.

Two initiatives were launched to improve the quality of construction in Singapore. First, we needed good construction workers. Since most of the construction workers were foreigners, there had to be a selection process to admit work pass holders. Only those who could pass a skill test would be given a work pass. At the same time, a CIDB Academy was set up to train local construction workers. The ITE construction department was transferred to the Academy to centralise all such training.

Another initiative was to set up CONQUAS to measure the quality of buildings in completed projects. CIDB in conjunction with major public agencies, leading industry professional bodies and organisations came up with a detailed template of criteria. The factors cover all aspects including structural, architectural and other services features in a building project. The CONQUAS score has been used for bonuses or penalties of construction companies tendering for public sector projects. The first session of the score was in place in 1989 during my tenure as Chairman.

Productivity was raised through the design of buildings by architects. Where possible modular designs have to be adopted so that construction prices can be lowered. Such practices were well-established especially in Europe. Another push for productivity is through construction contracts and contract labour. Under this system, payment should be made for finished products rather than for time spent. Encouraging the of use of prefabrication or precast components for buildings was also good for productivity. The quality of finished buildings is also better.

To help our construction companies to market their services overseas, a number of trade missions were organised by CIDB. The missions were relatively successful as a number of construction projects resulted.

For me, it was an informative and educational introduction to the construction sector. It proved to be useful for me when I had to take over Natsteel as its Chairman in 1989. Natsteel has substantial business in construction steel bars, prefabricated and precast concrete work.

Jurong Town Corporation (JTC)

In 1985, IF Tang was the Executive Chairman of EDB and Chairman of JTC. I got to know Mr Tang through the EDB, and he felt I could contribute to JTC and invited me to join him on the Board.

JTC is a statutory board under the Ministry of Trade and Industry. It is the government's developer and manager of industrial estates. During my tenure, it managed more than 40 estates covering more than 7,000 hectares.

In order to create recreational and social amenities, JTC developed a town center with shopping and commercial facilities, the Jurong Bird Park, the Chinese and Japanese Gardens and two golf courses.

During my tenure, JTC developed Jurong Island to house the petrochemical industry to diversify its economy. As the Singapore economy became more knowledge-based, JTC set up the Singapore Science Park to attract industries which produce high-value products. It also began developing one-north which houses Biopolis and Fusionopolis.

I retired from the JTC Board in 1995 after 10 years. It gave me good exposure to the work of the industrial estates which helped me to make good decisions in my future roles in companies.

Singapore International Foundation (SIF)

The Shanghai Municipality conferred the "Magnolia Silver Medal" on me for my work in forging friendship between Singapore and China.

In 2002, I was invited by Barry Desker Chairman of SIF to join the Board. Earlier, I had asked him to oversee the Institute of Defence and Strategic Sciences of NTU when SR Nathan stepped down to become President of Singapore.

The mission of SIF is to create better understanding between Singaporeans and world communities. It is funded mainly by MFA. Other than sharing my views on the various issues, my greatest contribution during my tenure of 2002–2014, was to initiate and be in charge of the YBA (Young Business Ambassadors) program with Shanghai from 2004.

The arrangement was to have 5 mid-level business professionals from each side to attach to the counterpart organisations for 10 weeks. They not only learned the business operations of the host organisations but also the culture and friendships of the staff in the receiving organisations. At the end of the 10 weeks attachments, a seminar would be arranged for the participants to share their experiences and views. It was found this program did bring about close friendships on both sides. The program lasted 10 years from 2004 to 2013. A formal celebration was held at the end of the program.

For my contribution to bring about friendships between Singapore and Shanghai, I was awarded the "Magnolia Award" by the Shanghai Municipal. A total of 100 YBAs from more than 60 companies have benefitted from the program.

Singapore China Foundation (SCF)

In 2004, PM Goh Chok Tong and his counterpart in China, Wen Jiabao, agreed to form the Singapore China Foundation to facilitate the exchange of officials from both countries. The discussions between the two PMs were held in Beijing.

On return, Mr Goh asked me to lead the program by being the Chairman of the Foundation. Initially, the Ministry assigned to support me was MFA as the objective was to improve relationships.

On discussion with MFA officials, it appeared that the simplest way to implement the exchange was to get officials on both sides to attend high level courses in the corresponding universities, i.e. NUS and Peking U. Supplementary visits to government departments could also be arranged. Due to this decision, MOE was then assigned to provide me with the support.

Every year, the Lee Kuan Yew School of Public Policy provided 4 scholarships to senior officials of China to attend its one-year Master of Public Policy program. In return 6 scholarships were provided to Singapore officials to attend a 2-month course in Public Policy in Peking U. The 2-month course in Peking U was normally attended by almost 60 senior officials from all provinces in China. The 6 from Singapore were the only foreigners allowed in the course.

From the feedback from our participants, the Chinese officials were free to discuss highly confidential matters. That was an eye-opener for Singapore participants who in turn gained unrivalled insights into China policies.

Since this was a residential course, all participants were housed at the university hostels, and they got to know each other well and remained close friends.

The course in Singapore was subsequently also offered by NTU's Managerial Economics program. Every March of each year, I went to Beijing to interview the Chinese candidates. March was a convenient time for the Chinese candidates because the senior officials gathered in Beijing for the yearly Assembly meetings.

I was rewarded with a lot of insights into China policies by interviewing and talking to the Chinese officials. Each year, I would have dinner in Beijing with our past scholars so that we could maintain links.

The scholars from China were selected by their Scholarship Commission (equivalent to our Public Service Commission). They came from various ministries as well as provinces. They were obviously very talented with a high level of linguistic skills — as we required them to be able to handle English to benefit from our courses.

I left the SCF in 2014 after 10 years. I learnt a lot about Chinese policies and practices.

Economic Planning, 1991, MTI

I participated in the economic planning exercise in 1991 as a member of the committee. The outcome was "The Strategic Economic plan: Towards a Developed Nation". The plan was to provide action plans for key industry clusters and the concept to strengthen economic cooperation in the region by partnering business and foreign governments to attract investors.

I chaired a sub-committee on "Education as a business" because EDB felt that this was a growth sector as can be seen by Australia and the UK. Members of this committee included some from private schools such as MDIS and the American school. Our deliberation came up with two recommendations:

(1) To encourage EDB to approve some of the best colleges in US, Europe and UK to set up joint ventures with the local colleges.
(2) There were too many private schools in Singapore at that time and some of them tie up with dubious colleges overseas. Many were also badly run, providing substandard courses.

The committee urged MOE to put in place a vetting system to eliminate the substandard schools so as to protect the good reputation of Singapore.

The Economic Plan was put up to the government for consideration. Our first recommendation on getting the EDB to promote Singapore as a destination for the top college overseas was approved. However, MOE was reluctant to implement our recommendation to vet the private schools. It was only in 2011 that the EduTrust was implemented to ensure the private schools and institution are properly run.

Land Transport Authority (LTA)

In 1999, the Government was full steam ahead and building many train lines. They thought that with my knowledge of SMRT, I would be very useful in LTA. I had to resign from SMRT as there would be a conflict if I were on the board of both the operator and the regulator.

The chairman was Michael Lim who had just retired from PWC. Another very well-known member was Dr Tan Cheng Bock. Both of us were in the EXCO, and as a result had to travel to many parts of the world to look at the operations and how different lines were managed. We looked at good examples as well as bad ones. During that time, I got to know Dr Tan very well. I had already met him in RI, as he came in one year after me and was Goh Chok Tong's classmate.

As we had many lines to construct, decisions came fast and furious — this was the most active board I had ever been on. The EXCO met fortnightly and we made decisions on millions of dollars — construction, parcels of land for stations and other matters. We had to meet the contractors and made sure they could do the job. We managed to track some very good ones for the Downtown line.

We were impressed with the line in Germany, Munich, because it was very smooth. You could hardly feel any vibrations nor hear any sound when you travel. We also thought quite highly of the Washington line.

What truly impressed us however was the Japanese MRT system. Despite the fact that the trains were crowded all the time, they ran well and were always on time. They counted the seconds and if a driver was out by 5 seconds he was considered late.

The whole system was so precise that you could stand on a specific spot at the station, and the train would pull up right to you.

We went all over the world to look at trains and tracks — here we are in Paris, the test track for Singapore's Circle Line.

Such precision and discipline impressed me because I was the chair of the audit committee.

I was also the chair of the R&D committee because we decided to put a significant amount of money for R&D. The committee gave out grants to external parties such as universities, institutes, research institutes. We marked out quite a large sum — today it is parked with The Rail Academy which I chair.

I retired from LTA board after 9 years — long enough since I always made it a policy not to stay too long on any board. I did feel that I stayed too long on some boards but it could not be helped.

When I retired, Michael Lim said that LTA must make use of my knowledge. When they formed the LTA academy, I was the inaugural chairman.

LTA academy was formed to do two things.

First, it was to look at policy, especially looking at how policies were made in other countries. If you were to look at the membership, the LTA academy has many international experts, so that we could use the knowledge of other operators, whether in UK, US or India.

We learnt how to tweak the policies to suit our own requirements and took advantage of the best of the transport policies from other countries.

Second, our LTA academy trained a lot of professionals especially from the ASEAN region. They came to LTA academy because we were quite well-known, not only for our policies but also for our strong implementation.

Many professionals were curious about how we deal with real life problems such as road congestion. We had Electronic Road Pricing which they found very enlightening. ERP 2.0 is coming soon, where we use satellites and do away with all the gantries.

Nowadays, roads which have gantries are regulated but some clever drivers just avoid the main road and slip into side roads. These side roads end up being more congested than the main roads. Once you have an ERP 2.0 on satellite, it will be just based on road usage. It is certainly more efficient than gantries, but I don't know whether this would be welcomed by the drivers.

The training of these overseas personnel was done at the academy itself. LTA officers, experienced staff and a few professors taught the courses. Not many people know but UniSIM and LTA Academy actually had a Master's course in transportation. Our professors in UniSIM, now SUSS, were roped in to give some lectures.

On the Board of the LTA academy, we had very well-known experts from England, US, China and India. All these were respected in their own fields. We had an annual meeting on specific topics, and each came prepared. We offered some secretarial support and transport allowance.

This was very little for the enormous contribution they made — but they all offered their time and expertise because they were deeply interested in the issue.

We had a visitor's centre very much like a Science Centre on transport. There were streams of school children coming in, learning about transport, and being quizzed about what they had seen and experienced.

The centre is in Little India, within the LTA campus and is a useful educational tool. You can see how our transport system started, what are things to watch out for, how to cross the streets, things like that. All these are very good for our young children.

I was Chairman of the LTA Academy from 2008 to 2014. Then, I thought I would have a bit of a rest. Two years later, Minister Khaw Boon Wan invited me for a lunch. As I have learnt, there is no such thing as a free lunch.

Singapore Rail Academy (SGRA)

Mr Khaw Boon Wan, Minister of Transport, appointed me as Chairman of SGRA in 2016.

The SGRA is a sub-group within LTA formed with the aim of professionalising the local railway workforce by collaborating with the rail operators, educational institutions and other key industry partners.

We train rail engineers and technicians with new skills. SGRA also facilitates applied rail research and technology applications with research institutes and IHLs. Eventually, it was to become a focal point for accreditation and certification of proficiencies for rail professionals.

In 2017, a small team was built up and together with staff at LTA mapped out the programs for the next few years. A board was formed with members from the university, polytechnics, senior staff of LTA, rail professionals and the two MRT operators to oversee the work of the SGRA management.

Mr Looi Teik Soon, a senior staff of LTA was appointed the Executive Director to look after the daily management of the Academy. SGRA could be relocated to the proposed integrated train testing centre (ITTC) currently being built at the site of former Raffles Country Club.

In front of Zhan Tien You, the first Railway Engineer of China.

I think this creates perfect synergy as the academy's training and research activities can make use of the testing facilities as well. In time to come, SGRA's roles could expand in order to meet the future demands of a convenient, well-connected, inclusive and fast land transport system.

According to the Land Transport Master Plan 2040, LTA will continue investing in public transport to make journeys more seamless and convenient. There will be the expansion of the rail network, including new stations and rail lines.

All these developments imply that there will be a future need for more manpower especially competent engineers and technicians to cope with the extra work. SGRA has a crucial role to ensure the skills of the technical manpower are available.

Over the years, I have led a number of teams of SGRA and LTA staff to visit training centres overseas like the UK, Germany, France, Hong Kong, Japan and South Korea. The learning experience would be useful to ensure we have a good knowledge of the requirements of competent training.

Each country has its own way of training. One important lesson for us is that in all of them, IT, VR, and Augmented Reality are important aids in training. Online continuous learning is also a strong feature.

Council of Presidential Advisors (CPA)

In my time, the Council had 6 members. They have enlarged it to 8 now. Out of the 6, 2 were nominees of the presidents, 2 nominees of prime minister, one nominee of Chief Justice and one nominee of chairman of PSC.

I was instrumental in nominating SR Nathan to be the president in 1999. In 2004, there was a vacant position. So, he decided to nominate me.

This was in 2004 and he had been president since 1999. The chairman was JY Pillay who took over from former National Development Minister Lim Kim San.

We visited many institutions of the state. For example, Ministry of Finance, Ministry of Trade & Industry, Monetary Authority of Singapore, Government Investment Corporation and the Housing and Development Board.

We had to visit all the institutions to make sure that they are healthy and their financial state are strong because they are the reserve of the country.

This could take quite a bit of time. For example, when we visited JTC, it took a whole afternoon. Occasionally, the president called for meetings to discuss important matters to do with the state of the economy.

It was during my term that we changed the rule for the use of Net Investment Returns Contribution (NIRC) At one time, the whole of NIRC was put into the reserves but we decided that it should be half.

The President would join these visits. Each of the organizations will brief us on the latest developments, their current problems, and their projections.

We got to know all the institutions well — especially their budgets. If there was a need to withdraw money from the reserve, we would make our recommendations. We were also involved in the appointment of all the key officers of GIC, Temasek Holdings, the PSC Chairman, the Chief Justice and the Chief of Defence Force.

All these have to go through the council.

The duties of the council are onerous especially when it came to the appointment of key holders, as well as decisions on the use of reserves.

The President cannot lightly disagree with the recommendations of the council. In order to disagree, he would have to go to the Parliament for two-third majority to overrule the council. That is why while we have tremendous power, but also tremendous responsibility.

We had to be sure that what we had decided and done was right.

During my term, our voice was unanimous. Sometimes we faced situations that were delicate or highly unusual but we had always been unanimous in our decisions.

I served one term of 6 years. As events turned out, I only served President Nathan.

Singapore Quality Awards (SQA)

I was roped in by Mr Lim Boon Heng in 1993 to start the Quality Award movement. At that time, he was the Chairman of the National Productivity Board and Minister for Trade and Industry. His reason for getting me to chair the SQA Governing Council was because he thought that I would understand productivity well as I was an engineer. Second, I was then President of NTU and considered to be impartial by companies and organisations. A senior officer from NPB Freddy Soon was assigned to assist me.

For a start, I thought it would be useful to learn from similar organisations overseas. There was the most prestigious award called Malcolm Baldrige National Quality Award which recognises US organisations in the business and other sectors for performance excellence.

The awards are given by the President of the United States himself. It was instituted in 1988 and named after the United States Secretary of Commerce during the Reagan administration. I flew to Washington to meet up with the Head of the National Institute of Standards and Technology (NIST). He was kind enough to give me the dos and don'ts of starting such an award.

Next, I went to Brussels to find out more on the EQA (European Quality Award) started by the European Foundation for Quality Management. It was started in 1992 (two years before SQA) and companies in Europe are awarded for Total Quality Management. There are 4 categories, 2 for large and small private enterprises and 2 for large and small public organisations. The criteria adopted were very similar to the Malcolm Baldrige Awards.

The last destination of my fact-finding missions was to Tokyo to find out about the Deming Prize which was first awarded in 1951 for organisations with the best Total Quality Management. The sponsor was the Japanese Union of Scientists and Engineers (JUSE). There is also a prize for an individual who has contributed most to the quality movement.

From the studies, I compiled a list of good practices which were useful to Singapore at that time. Since the initiative was from the National Productivity Board, "productivity" of the company was stressed as an important criterion. Later, other criteria such as "people" and "innovation" were included.

Over the next 20 years from 1994, we reviewed the criteria 5 times to make sure they remained relevant. For example, the ability of the company to thrive in digital transformation is stressed to make sure it is able to deal with the disruptive technology. We also instituted Quality Class Certifications for companies which were not ready yet for the SQA Award, but have done well enough to be given recognition in certain aspects of its management.

To date, about 2,000 organisations have benefitted from the Business Excellence initiative adopted by the SQA committees and facilitated by ESG (Enterprises Singapore). Within the BE are the development tools to support companies for improvement.

At the 25th year of the SQA in 2019, the number of BE (Business Excellence) awards was 121 — won by 74 organisations with 61 the top SQA Award. Companies are now focussing on BE as enterprise development instead of bidding for recognition with certification or awards. We have decided to end the SQA Awards as from 2020 because they have served the original purpose of encouraging companies to excel.

Chapter 7

Industries

NSL (NatSteel, National Iron & Steel Mill)

NSL has a long history as a listed company. I was appointed as a Director of the company NISM in 1988 mainly due to the fact I was Chairman of the Construction Industry Development Board (CIDB). NISM at that time was mainly in the steel mill business and construction by its subsidiary company Eastern Pretech. I was therefore able to advise them in their business.

NISM was set up in 1961 as part of Singapore's development plans. It was to supply steel to Singapore's construction industry in conjunction with the HDB housing plan. EDB supported it by taking a 20% stake in the company. It was the first factory in the region to roll steel bars out from scrap metal from local and overseas sources.

In the early years, the demand for steel from Singapore's construction industry absorbed NISM's entire output, but exports became significant from 1973.

Another early investment was with Finnish company Partek to produce prefabricated concrete parts for quicker construction. The Singapore operation of Partek was named Eastern Pretech.

In May 1990, NISM changed its name to Natsteel. In 2003, 98 Holdings led by Ong Beng Seng took control over Natsteel. Eventually, it owned 81% of the share of Natsteel.

In 2004, the steel division of Natsteel was taken over by Tata Steel, as a consequence the company's name was changed to NSL. The remaining businesses are construction under the name of Eastern Pretech Singapore and Eastern Pretech Malaysia and environmental under the name of NSL Oilchem.

I was Chairman from 1989 until the present. Over almost 30 years, I saw and managed the changes.

The first CEO was Ang Kong Hwa who was a good business man and entrepreneur. Although he was formally trained as an economist, he is tech-savvy. He could operate the steel mill in one breath and work with computers in the next. He retired to become Chairman of Sembawang Group.

The next was Gan Kim Yong, the current Health Minister who joined the Natsteel Group in the late 80s. He went into politics in 2011. Dr Josephine Kwa, Oo Soon Hee and Low Chin Nam were the later CEOs I worked with.

I enjoyed running NSL as it provided me with a range of experiences across many industries. We even got into owning resorts, such as a golf course which was eventually sold. We still own the Raffles Marina. We had ventures overseas such as a steel mill in Brazil, a petrochemical company in Thailand and a construction company in Dubai.

Singapore Press Holdings (SPH)

In 2004, Mr Lim Kim San was the advisor to SPH. He had stepped down as the chairman of SPH, but was still very much a man trusted by the highest leaders in Government. He had a special power of knowing just what is required in any circumstance.

He got me in as the deputy chairman. The chairman Lim Chin Beng was already 71. I supposed he thought that I might succeed the chairman once I knew my way around.

Mr Lim thought that I had some knowledge about Singapore. I had started NTU and it was doing well. I was on various boards. Knowing what Singapore was like, he thought, was good for SPH. At the same time, in 2004, I was called into the Council of Presidential Advisers. So, we thought it was a good time for me to be in SPH.

SPH is very different from all the other companies I had experience with.

Over the years, I have the chance to have many lunches with Mr Lee Kuan Yew, he always told me that any government should prioritise three things.

Presenting MM Lee with a birthday gift on his 88th birthday with SPH Senior Executive Vice President Robin Hu.

First — the economy which is obvious because if the economy is not doing well, you would have problems with employment. Next the military, because in international politics, power counts.

He had a "poisoned shrimp" theory — you must behave like a poisoned shrimp so that anyone who tries to eat you will pay a very high cost. If you could withstand any external threats for 24 hours or maybe even 48 hours, the friendly countries will come to your rescue. Hence, he said the military was crucial in providing this credible threat.

Third, the media. I can see his reasons. You do not want the population to be influenced by views which are unfriendly or perhaps not very constructive. Mr Lee was very careful with the media.

He told me that Straits Times is not just a newspaper but a record of the country. When you looked back to see what Singapore was like, you look at the ST archives. That is actually the official record of the country.

Therefore, there are two things that he makes sure of for SPH. SPH has an allocation of "golden shares" owned by selected parties. The directors of SPH were also carefully chosen.

His philosophy is that the media has to support the government insofar as it advances the public interest. Of course, we may differ in what exactly this means. Over the years, we have had occasions when he was not exactly pleased in what we had written and we got very timely feedback.

For example, when former President Devan Nair passed away, we thought it was right to give him some prominent publicity especially given his work on NTUC. However, Mr Lee felt we had overdone it. This was a case which many ministers felt strongly about as well — so we received many queries from ministers too. Ministers also write in if they had made a speech and we did not report it. Sometimes the editors are pulled into the ministry to answer questions about our editorial decisions.

In 2005, Tony Tan became the chairman. When he got elected as president in 2011, his first choice was to ask me to succeed as chairman. But I was apprehensive, so I cast around for a better candidate.

The PM suggested Lee Boon Yang. As the former Minister for Communications and Information and the Arts (MICA), he would know the landscape well. Being the chairman of the nominating committee, it was my job to approach Lee Boon Yang.

He was at the time chairman of Keppel Corp so I went to his office to make the offer. He declined diplomatically, saying that Keppel kept him very busy. However, he subsequently agreed.

I thought this was instructive — clearly, it was not an easy job and even the former minister thought it was a hot seat.

My main contribution to SPH was perhaps during halfway through my term — we were essentially a media company but I thought that we should diversify into property. My rationale was that there would be times where the papers would not be doing well in advertising. At least property would prop it up, and would moderate the business cycle.

When we told MICA, they were unhappy as they said we were a media company with no business venturing into property. Fortunately, Tony Tan came in about that time. He too thought that property was something useful to moderate the business cycles.

So, we went into property and SPH REITs became a listed company. Events proved me right — today, SPH's profits come because of its property portfolio and not because of its press.

When Lee Boon Yang came in, I told him that my 9 years were up and that I wanted to step down. He asked if I could delay for a year so

I left SPH in 2014 after 10 years (picture, SPH Farewell Dinner with Chairman Lee Boon Yang, in 2015).

that people would not think that the deputy chairman quit because he could not get along with the chairman. So I stayed for a year till 2014.

It had been an eventful 10 years.

When I first joined SPH, there was only one TV station. The Government later decided to have another. The Board was not keen because we knew that Singapore could not accommodate 2 TV stations given our small size. Running a TV station is much more expensive than running a newspaper.

Despite our reservations, we still did it because the spirit of the times was to believe in competition. Similarly, I had always thought that it was not very sensible to have 2 MRT companies in Singapore when the business was so small. Hong Kong, which is double our size, has only one.

Today, we are not so seized with the idea of competition for its own sake. We no longer have a TV station. The SMRT has been nationalised. The newspaper TODAY is completely online.

Another memorable episode was the day when political detainee Mas Selamat escaped on Feb 27, 2008. On the same day, there was a rumour on the passing of Dr Goh Keng Swee. Coincidentally, I was at the United Overseas Bank's board meeting with Mr Wee Cho Yaw when the Prime Minister's Office phoned him to ask about the rumour as it is well-known that Mr Wee was close to Dr Goh.

Coincidentally, an hour later, Mrs Goh Keng Swee also phoned Mr Wee. Mrs Goh was close to Mr Wee, and phoned him occasionally. He asked her if Dr Goh was still alright and she said he was fine. That was the end of it — or so I thought.

On the same night, at that time about 11pm or so, one of the senior editors called me and wanted me to give him a few anecdotes about Dr Goh. It was strange so I asked him why he was doing this, and he replied that Dr Goh had passed away. The newspapers had already laid out the page, with a big square on the front page for Dr Goh's face.

I told him to check. Of course, it turned out to be false.

In the end, the big square that was supposed to be Dr Goh's photograph was replaced by Mas Selamat's photo. If you look at The Straits Times, his photo was prominent on the front page.

Dr Goh passed away two years later in 2010, aged 92.

Formula One (F1)

I have been very passionate about F1 partly because of my PhD at Cambridge. My research work there was on high speed aerodynamics.

When a vehicle travels very fast through air, a thin layer of air is attached to the surface of the vehicle known as the turbulent boundary layer. This layer is due to the friction generated which in turn can slow down the vehicle. F1 cars are designed such that there is minimum drag caused by the air. This is called streamlining the car.

Designing this streamlined car is technically quite complex. Usually the first step is to produce a computer model and check the air flows by computer simulation. The design is then tested by putting a solid model in a wind tunnel and taking appropriate measurements.

F1 cars are also fitted with "wings" which provide the downward force so that the wheels can generate the drive. Of course, it has to be a balance between the drive and the drag. So, the design of the wings is very important.

F1 races were started in 1950 in London. It consists of a number of races in different countries. For 2020, the number is 21. Of course, the actual number depends on whether the Covid-19 will become serious in the country concerned. The word "Formula" refers to the set of rules to which all participants must conform.

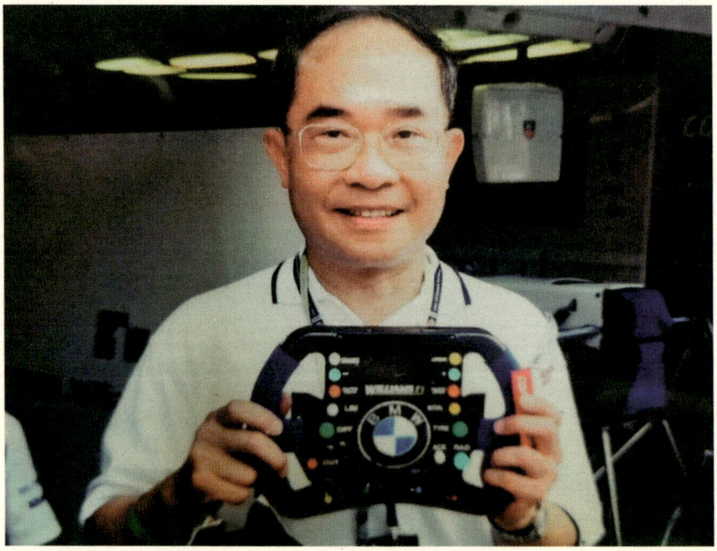

Speed has always fascinated me — from airplanes to Formula 1 cars. Here, you see the complex steering wheel an F1 driver has to deal with (2008).

F1 races are very popular in Europe. It only came to Asia in the late 80s with Japan 1987, Malaysia 1989, China 2004, and Singapore 2008.

Since I was involved in the planning for the race in Singapore, I can give more details. The person who was instrumental in bringing F1 to Singapore was Mr Ong Beng Seng. He was a personal friend of Bernice Ecclestone who was then the CEO of the organising company. After some negotiation it was agreed that Singapore would have the first night race in the city from 2008.

The Singapore government would fund two thirds of the cost of running and the rest would come through ticket sale and commercial activities. It is believed the yearly cost of running the F1 in Singapore is in the region of $150 million.

In 2007, one year before the race, I was asked to meet the Working Committee to give advice on technical issues. Mr Ong trusted me because of my prior interest in F1, having attended about 10 live races prior to 2007. Most of the issues could be resolved but some were not feasible because of the high cost.

Being a night race, the total length of race lighting of 6km was fitted with 1,500 light projectors.

In order to plan for emergency power in the event of failure of the electricity grid, it is also fitted with standby generators giving adequate power to light all the projectors. Fortunately to date, no power failure has occurred.

One of the attractive features of running a night race in Singapore is because it is run in the city. With all the lighting of the buildings, the skyline is very beautiful. The race is broadcasted all over the world, beaming across the night skies of Singapore.

In 2008, I attended the first F1 race in Singapore — the roar of race — these cars are more alike to fighter jets than cars — was incredible. I have met a few racers — Ralf Schumacher and Juan Pablo Montoya — and once while I was in Canada for the Grand Prix in Montreal, I got into a cab who took me for a spin on the race track.

The bends were so sharp we got thrown about in the back. It was quite a ride.

With F1 driver Juan Pablo Montoya in 2008. He is one of the top drivers of all time.

With former Prime Minister Dr Mahathir at F1 in 2009.

Keppel Corporation

I was put to join the board in 1982 which was recommended by the Ministry of Finance. At that time, the chairman of Keppel was George Bogaars and I knew him in two ways. One was as a Perm Sec at Ministry of Finance — I had a lot of dealings with him then because I had to start NTI. As a Permanent Secretary he was often dour and always very strict. I suppose it was his job to make sure that the funds were carefully managed.

When I joined the Keppel Board, I became a part of his team and we worked well together. He seemed to be a different person, very joyful. We both wanted to make Keppel do well. The fact that I was an engineer by training meant this was right up my alley.

In addition, he thought that I would be a useful person to start the audit committee in Keppel. At that time, we were the first and only audit committee in Singapore among the listed companies.

It was not a requirement to have an audit committee. In fact, I had the opportunity to meet up with the Auditor General of Singapore to understand the remit of such a committee. He was kind enough to hold a lunch and was quite frank in telling me the do's and don'ts. At the time, Pricewaterhouse was our auditors.

There were two reasons why we set up an audit committee. One, as a precautionary measure. Second, Keppel was (and still is today) a complex company. At that time, they owned many ships and shipping was not doing well. They also owned many properties — Ocean Tower at Raffles Place among others.

We were unsure whether our finance department was up to the mark. As it turned out, the audit committee was certainly set up at the right time because recession came about 1985. We were able to make sure that whatever we have was something which was defensible.

At that time, we were in a really bad state — our balance sheet was negative. Property prices were falling and shipping did very badly. We had bought many ships and all were idle.

Fortunately, the recession did not last very long. We were able to sell off a number of ships but at a great loss. In fact, we wanted to sell the Ocean Tower but luckily, did not — there was a bidder for the property but when we checked its profile it turned out to be a two-bit company. Therefore, we decided not to sell.

When the things took a turn for the better, we built an additional tower. So instead of selling, it became a twin tower, instead of folding, we expanded.

Dr Goh was then in charge of finance. He was not happy with the performance of George Bogaars and replaced him with Sim Kee Boon who came in after retiring as a head of civil service.

I spent a full day to bring Mr Sim up to speed on Keppel and outlined what we intended to do for the future. I briefed him personally, one-on-one because I chaired the audit committee so I knew the financial circumstances. Although he was the head of civil service, he was also a very good businessman.

He led Keppel for a long time.

Mr Sim managed to turn things around. Our economy also picked up after 1985.

Mr Bogaars was Chairman at a difficult time — some of the things he did, I would not fault him for. In a large business, you have to take certain risks. Otherwise, you would not grow.

Mr Sim knew what was good for business. He was able to come out with strategies or moves that benefitted Keppel. For example, he went into property, Keppel Land and we even bought over a bank and called it Keppel Bank.

After he left the civil service, Mr Sim Kee Boon (fifth from left) became the Chairman of the Board of Keppel. (I am second from right).

I thought that was wise of him to diversify into other fields. That was how we grew.

Later on, Lim Chee Onn took over.

Lim Chee Onn has been working at Keppel for some time. He used to be an MP, and headed NTUC. However, things did not work out so he joined Keppel as a senior employee after dropping out of politics.

I got to know him at board meetings. By the time he became chairman, we already had a long working relationship and I think he felt I was someone he could trust. I chaired the audit committee, the risk committee and the remuneration committee.

By 2003, I decided that I had been on the board long enough.

Keppel Technology Advisory Panel (KTAP)

When I retired from Keppel Corp Board in 2003 after 21 years, Mr Lim Chee Onn, then Chairman, invited me to Chair a new panel (KTAP) for Keppel. I chaired the Panel for 8 years from 2004 to 2011.

Keppel management were convinced that going forward, Keppel could not take success for granted by doing the same things over the long term. The answer to how best to seek and sustain growth for the Group was through the use of technology.

For Keppel to have access to this knowledge, there was a need for an International Advisory Panel which could include eminent scientists, business leaders and industry experts from across the world.

During my tenure, the eminent members were:

(1) Professor James Leckie, Professor in Environmental Engineering, Stanford University
(2) Sir Eric Ash, Former Rector, Imperial College, Vice President of Royal Society
(3) Sven Bang Ullring, Chairman, Det Norske Veritas
(4) Dr Brian Clark, Schlumberger Fellow
(5) Dr Malcolm Sharples, Vice President, American Bureau of Shipping
(6) Prof Minoo Patel, Dean of Engineering, Cranfield University
(7) Prof Thomas Curtis, Professor of Environmental Engineering, Newcastle University, England
(8) Dr Yeo Ning Hong, Former Cabinet Minister, PhD (Cambridge) in Chemistry

Over the years, KTAP members have contributed to a broad range of ideas and development for Keppel. The areas include drilling and production technology, offshore wind, coal gasification, waste-to-energy, as well as potentially disruptive technologies.

As Mr Lim Chee Onn put it in the inaugural KTAP meeting on 20 May 2004:

"We are growth-oriented, and the areas that we are focusing our main efforts on are in three key areas. The first is marine and offshore, the second is in real estate, and the third is classified broadly as infrastructure, which in itself has three businesses — primarily in network engineering, power generation, as well as in environmental engineering."

"When we selected these three-core businesses, there were a few key factors which determined their selection. The first is quite obvious: that whatever core business we select we ought to have the core competencies in that business. The second, which could be a bit controversial, is how we can determine the potential growth of that particular business. In short, we do not focus on the business only because we can do it and know a lot about the business but it must have the growth potential."

"The growth forward, if I may put it in a nutshell in our view, would depend on technology, because we believe, and I personally endorse this — technology would be the differentiating factor, not only in costs but in the products and in some instances packaging. This is where the formation of KTAP was first mooted. We thought that it would be marvellous if we had a panel of eminent scientists, technologists, businessmen to be able to help and guide us as we try to look for the road map to move forward."

KTAP met twice a year with key members of Keppel's Board members, and senior management.

Distinguished guest speakers were often invited to these meetings to share the latest developments in their respective fields. Apart from meetings, frequent discussions were coordinated by the Secretariat through email on topical issues such as shale oil, nuclear energy and sub-sea related developments.

I did ask Mr Lim Chee Onn the question of the roles of the Keppel Board and KTAP. His answer was concise but clear: "The Board has to look after the day-to-day operations as well as year-to-year progress of the company. KTAP has to think of the relevance of Keppel businesses in ten years' time."

In 2012, I stepped down from KTAP.

UOB

At the turn of the millennium, Mr Wee Cho Yaw invited me to be on the board. I knew him before that as the former chairman of the Nanyang University Council, and from the interactions with the Nanyang University Council and the Nanyang University Alumni.

He invited me to be on the board which I cherished it because it was my chance to be exposed to a bank and the financial world. UOB is also one of the largest banks in Singapore.

When I went onto UOB board, I worked very well with the chairman, Mr Wee. As a result, I was in every subcommittee of UOB which was unusual. I was the only one in that situation among the directors.

As chairman of audit committee, I had to meet MAS at least once a year, where they discussed the strengths and weaknesses of each of the banks. That gave me exposure and idea as to how MAS works. That was very useful with my dealings in other boards. In the end, Singapore's survival depends on the world of commerce and trade.

We took over OUB in the first year I joined.

It was uncommon for a bank to take over another bank. At that time there were 4 banks, but the Government felt that Singapore financial business was not big enough to have four.

The smallest bank then was OUB.

The Government had at first suggested that DBS should take over OUB, because DBS was the biggest out of the four. A director of DBS Fock Siew Wah was assigned to speak to OUB. The chairman who was the late, Lien Ying Chow, did not take the news well.

UOB also made it known that it was interested in OUB. In the process of getting shareholders' agreement for the takeover, DBS painted a bad picture of us. UOB had no choice but to bring a legal case against DBS, who apologized and paid $2 million in damages.

In June 2001, DBS announced a takeover of S$9.4 billion. UOB made a competing bid of S$10 billion. UOB's bid was a friendly one and on September 2001, we had 90% acceptance from shareholders and succeeding in delisting OUB.

That was one interesting episode during my 10 years. Another was when I chaired the committee to divest.

There was a new banking act which said that any one bank could not own more than 10% of non-financial entities such as hotels. At the time, we had more than 10% interest in OUE and United Overseas Land (UOL). I had to help divest a lot of our non-financial interest. The MAS saw it as a way to reduce risk.

In 2006, we sold OUE to Stephen Riady for a big sum of around $2 billion. We were a bit apprehensive on the day of settling. Luckily, he came with a cheque from a trusted Austrian bank.

I got on well with Mr Wee and his son, Wee Ee Cheong. Having a father who cast a very long shadow must not have been easy for him. But in fact, Ee Chong is a very capable banker.

UOB (China)

When I reached 75, I could not be on the board of bank UOB as this was against the MAS Banking Act. I retired from UOB Singapore.

Mr Wee told me to remain in the UOB China as a director. Hence, up to today, I am still involved with UOB China.

Wearnes Group (WBL)

WBL has a long history from 1912 when it was listed as Wearnes Brother Limited, a public company. It had a wide range of business interests, including a commercial air service serving Singapore, Kuala Lumpur and Penang in 1937. It was called the Wearnes Air Services.

In 1985, Mr IF Tang, the Chairman, invited me to join the Board as a Director. At that time, the main business of Wearnes was in the automobile trade. It had in its portfolios, the Ford cars, Rolls-Royces, Jaguar and Bentley. There was then the plan to diversify into technology and manufacturing services, Mr Tang thought that being an engineer I would be able to contribute to Wearnes.

The first venture into technology was to manufacture its own brand of desk-top computers, called Wearnes Computer. At this time, it consisted of just purchasing all the parts and assembling them into a complete computer. This process was actually making Wearnes a system integrator. As the business progressed, the company started to make the components such as the motherboard, chipsets and even disc-drives.

A subsidiary company called Wearnes Technology was set up to manufacture such components when the business expanded. The manufacturing activities also expanded overseas such as Thailand and China.

We improved our know-how and technology very quickly through a number of strategic acquisitions. The most well-known was the Advanced Logic Research in California which gave Wearnes the edge needed in computer technology. This acquisition strategy was started in 1985, the year I joined Wearnes.

In fact, in 1989 Wearnes was among Wall Street Journal's top 66 companies of the future for its venture into high technology.

A significant diversification of Wearnes was its involvement in property developments. Although due to its long history, it had already substantial ownership of properties in Singapore and Malaysia which it had acquired for its automobile trade. From 1989, it began an aggressive strategy to develop properties in China. It started off in Shenzhen in 1989 with the building of a multi-storey workplace for high technology ventures.

Then it went on with substantial property developments in other cities such as Chengdu, Suzhou, Shanghai and Shenyang. Most of these developments are for high quality condominiums and are considered very successful by the Chinese. Due to its presence in China, Wearnes was renamed WBL corporation in 1994. It was felt that the name "Wearnes" was too foreign for the Chinese.

I retired from WBL in 2011. It was subsequently acquired by United Engineers.

MFS Technology

MFS was a subsidiary company of Wearnes and specialised in electrical and electronic subcontracting work. It was involved in PCB (printed circuit board), chipsets and electronic components for computers, handphones and electrical parts of automobiles.

Its headquarters was in Singapore and had manufacturing plants in Malacca and Changsha, Hunan.

MFS was not doing well for some time and I was instructed to chair the Board in 2007 to restructure the company and make it profitable. The significant steps I had to take were to change the CEO, put in place a high-powered business development VP and restructure the management of the Changsha plant. By 2009 we were profitable again.

Two interesting events during my tenure were associated with our presence in China. Xiaomi, the Chinese handphone maker used MFS Tech to supply the motherboard chipset for its first version of handphone. It was very Apple-like in its appearance. Being the Chairman of MFS, I was given one of the first production of handphones as a souvenir.

I did visit our Changsha plant in 2010. It was well maintained and up to date in its facilities. I also visited the birthplace of Mao Zedong. It was a village with a beautiful setting of a large lake in front of a house which was sparely furnished. It gave me a glimpse of his simple lifestyle.

I retired from MFS Tech in 2013 when it was acquired by a Private Equity Company.

Singapore Automotive Engineering

When I was still the chairman in the years 1985 to 1989, it was called Singapore Automotive Engineering. When it became listed, it became ST Kinetics. I was there before it went public.

When I was in SAE, our main portfolio was looking after the military vehicles, the tanks, armoured personal carrier and even the army truck that carried people. We did the maintenance and were also quite active in R&D.

We tried to make our tanks and our armoured personal carrier better. For example, we took the tank body from the Indian army and improved the inside with many new parts and functions.

We did not have to assemble the tank. We added enhancements such as an automatic drive and changes to firing power. We looked after armoured personal carriers.

At one time, we were experimenting with an amphibious vehicle — a tank that could float on water and then go on land. We did all these things and basically looked after the hardware of the military in terms of transportation.

We did some confidential work which not many people were aware of, including making equipment with attack functionalities such as the Century tank. Every once in a while, we brought these out on moonless nights, to test them out.

That was the military part, but we had to do the civilian and commercial part too, so as to fund these military expenses.

Singapore Automobile Engineering was a good repair centre and maintenance centre of cars and commercial vehicles. We even ran a taxi company City Cab to generate the income. I do not know who is looking after the Opel brand now but we used to be the main distributor of Opel from Germany. As a consequence, I owned an Opel car at a point in time, just to show the confidence in our maintenance workshop. City Cab, which was quite profitable, eventually merged with ComfortDelgro.

We had visits from Yeo Ning Hong who was then Defence Minister. We had put in a lot of effort in the enhancement of AMX 13, and had a ceremony to launch the new and improved version, with the Chief of Defence Force Winston Choo and I. I have a memorial copper which is still hanging in my office.

SMRT

The construction of the MRT line was under the proper MRTC, chaired by Michael Fam.

Michael Fam is a very shrewd businessman. The first phase of construction of the North-South line came in under the $2 billion budget which was very unusual. During those times, construction nearly always overran.

It was partly due to the entrepreneurial spirit of Mr Fam.

The construction line was great success. Every station has a clock to tell the time. If it were any other chairman, we would have asked to tender for the procurement of 100 clocks but he reversed it and asked which clock company wanted to advertise their clock in the stations — and how much they were willing to pay MRTC for the privilege.

Instead of expenditure, it became an income. Such sponsorship and advertisements were part of a very shrewd business plan.

The first line was supposed to be ready in 1987 for the first run in November. The SMRT was incorporated with 100% funding from Temasek Holdings. At the time, the Minister of Communications was Yeo Ning Hong, who was my personal friend from Cambridge. He came to Cambridge after I did and we crossed over so we often had dinners together. My wife and I knew him and his wife.

In 1987, he called me and asked me to serve on the SMRT board so I became a director of SMRT. I was quite active, as I was not only on the EXCO but also chaired the audit committee because of my experience in Keppel.

At the time, Michael Fam chaired MRTC in the controlling and construction, Fock Siew Wah oversaw the operations and the MD was

Lim Leong Geok. The initial operations were quite good and since the funding came from Temasek Holdings, we had some surplus.

12 years after the first run of our trains, Temasek decided to list the company to recoup some of its investments. In 2000, it was listed as SMRT Corporation and Temasek Holdings sold 33% of the shares. But it retained control via the "golden shares". They did not need to have a majority share because no one could take over SMRT unless they held these golden shares.

I was not keen on the listing — if we listed it, thousands of shareholders would want the company to give strong returns. In order for it to be profitable, I was worried that certain things would bow to the economic imperative. Maintenance was costly, so were replacement parts.

When Temasek Holdings was the owner, profitability was not a major consideration although we made sure that we were not running big deficits. We were providing a service to the people. That was how I saw it.

Still, because SMRT was not running a deficit (indeed we had some surplus), it was not difficult to list it especially in 1999. I think the investment public was quite keen on new shares.

The listing was oversubscribed and very successful. As a retiring director I was given some shares as a parting gift. As I moved on to LTA, my LTA colleagues pointed out that I could not own these shares because it was a conflict of interest. So, I sold them.

The running of a railroad is a very interesting endeavour. For any new line constructed, we had to make sure that there is viability in the passenger load. The timetable is justified according to passenger load. We estimated the lines according to new housing estates and based on HDB forecast of residence numbers. Looking ahead, however,

Singapore will eventually have a lot of lines crossing one another so it is no longer on the basis of passenger load.

Instead, we must work on the basis that every household must be within 10 minutes of walking distance to the station anywhere in Singapore. By 2030, the mileage per person will double.

The exact location of each station will depend on whether we can acquire vacant land and the ease of construction. For example, the Bukit Timah Downtown line had to constructed in such a way as to avoid the canal.

The timing of construction of the lines may also depend on wider circumstances. We planned during 1982 and 1983 that our first construction will be in 1987 but we brought forward construction to 1985 during the recession as we anticipated that contractors will be eager for contracts so they would bid low.

The tender for all equipment and design was mostly submitted by foreign companies as we did not have experience of tunnelling construction of lines. Our contractors still benefitted from being subcontractors. Even now, many projects such as the Thomson East Coast line have main contractors who are not Singaporean. They are either Chinese or South Korean but our contractors are the sub-contractors.

We made many study trips all over the whole world.

US, UK, Europe, Hong Kong, Taiwan, China. At the SMRT board and LTA board, we made it a point to travel to see how the trains run, what the problems were and what to look out for. We wanted to get the best practices all over the world.

The first phase of MRT in the 1980s took reference from the London Underground and the initial employees in MRTC were from London Underground. The government set up an independent organization called the Public Transport Council which decided on the fares.

The opening of the MRT in 1987 was witnessed by the Prime Minister, the Minister of Transport Yeo Ning Hong, Michael Fam and Fock Siew Wah. I kept a souvenir of the first ticket. I even had a plaque commemorating the first run of the MRT. Later I gave it to SMRT for their archives.

Soup Restaurant (SR)

Soup Restaurant was started in 1991 from a shophouse in Chinatown. After 29 years, it has grown to a restaurant chain with outlets across Singapore, Kuala Lumpur and Jakarta.

SR specializes in serving Chinese comfort food with its signature dish of Samsui Chicken. It was reported the recipe for the Samsui Ginger Chicken was handed down from the Samsui Woman in the 1940s.

It now has 15 outlets in Singapore, 1 in Kuala Lumpur and 1 in Jakarta. It was listed at the Singapore Stock Exchange in 2007 and I was invited to serve on the Board by the founders.

The four founders were my engineering students in the University of Singapore in the 1970s. When the company tried to list it at the SGX, they were told the company needs a director with a good corporate governance experience. That was my role.

As the Chairman, I worked not only with the Board, but also very closely with the Executive Directors who were the founders. I am still having monthly meetings to sort out some of the difficulties confronting F&B companies.

Recently, SR has done well in CSR work by having a central kitchen at Changi Prison to train inmates so that they would be employable on release from the prison. It also has a training centre for the disabled at the Enabling Village.

I am enjoying my experience of working for the SR which is very different from my other industrial companies.

Mott MacDonald, Singapore

Mott MacDonald is a large British engineering consulting company specializing in infrastructure development. It is well known for its involvement in the London Olympic in 2012 developing the facilities for the games.

In Singapore, its recent projects were with LTA and Singapore's future Terminal 5.

I was invited to be its advisor from 2004 to 2010. My main advice was in its approach to projects and the manpower required for implementation. It opened my eyes to how a British consulting engineering company operates in this region.

SAP (APJ)

SAP is a well-known accountancy firm started in Germany. Its software and accounting processes are used by almost all large companies in the world. APJ stands for Asia, Pacific and Japan. I was invited to be on the Advisory Board from 2014–2018.

SAP is now very much driven by IT. It is in the Cloud business as well as Real Time Accounting. It is a well-regarded company. Being on the board gives me good exposure on how such "aggressive" companies flourish.

Asian Association of Business Incubators (AABI)

I was asked to represent Singapore in the formation of AABI in 2002. At that time there were only 5 other countries/regions namely Japan,

China, South Korea, Taiwan and Hong Kong. Now it has grown to 19 country-members. The first meeting was held in Osaka, Japan and we laid down the objectives, missions and membership issues. The objective was to link business incubators in the region by the exchange of information and ideas.

We meet twice a year, the venues being rotated to each country.

From 2010, we initiated competitions for Incubators of the Year as well as Entrepreneur of the Year. They have been well supported. The panel of judges came from the founding country representatives. So, Singapore is a permanent member of the judging panel and the Board. Under its wing, is the Asian Association of Business Incubators Journal (AABIJ) which is well publicized and widely-read.

For my role in promoting AABI from the beginning, I was awarded the AABI Lifetime Achievement Award in 2018. Two others were also given such awards.

I received the AABI Lifetime Achievement Award in 2018.

A closer look at the AABI Award.

Exposure to AABI gives me a good idea how business incubators are promoted and seen in each country. They are different with different cultures and policies of the respective governments.

Chapter 8

Looking Forward

For the previous chapters, I have written on my past experiences and encounters with various people and organisations. They serve as historical records of what I went through.

For this chapter, I wish to project what the world will be like in the future, perhaps in 20 years' time. The nature of change will be unprecedented and would be difficult to imagine at this point of time. Predictions of such nature would also be difficult to substantiate.

More dramatic would be the *pace* of change. Even today, things we take for granted would have been inconceivable a decade ago. For example, the way we use our handphones for our daily routines would be unimaginable even five years ago. We can go around without our wallet but definitely not without our handphones. I use the MRT quite often and it is interesting that almost every person looks intently into their handphones.

Disruptions due to advancements in technology will be even more dramatic for businesses and industries. Many experts think that we are at a crucial moment in history when disruptions due to technology are unpredictable, fundamental, and revolutionary.

An appropriate term to describe this phenomenon is "disruptive innovation".

The main reason for this massive disruption is the adoption of wide-scale automation. Businesses are also faced with mounting pressure to decrease costs and increase productivity. That would set up a wave of changes.

There are two enablers which makes wide scale automation attractive and easy to implement. The first is the ease of acquiring big data. This is mainly due to the rise of computing power and availability of devices and sensors which capture that data in a trend called "the Internet of Things". The data itself is not powerful but what the data can be used for that counts. The data gives indication on how and what are useful automations.

The second enabler is "Artificial Intelligence". Our computers are smart and getting smarter all the time. The reason for this is the increasing pressure on time for operations of computers and the storage space for data is getting more efficient and cheaper as well. So, it is a matter of using the appropriate software to configure the computer to "shrink". The speed of this process makes automation easier and more effective.

The development for the businesses and industries is now called Industry 4.0 or Fourth Industrial Revolution. It is the "Digital transformation of industries and society". The First Industrial Revolution was the transition of muscle power to machine power. The Second, introduction of electricity and the development of the production line. The Third, computer revolution — development of the mainframe and Internet.

Due to all these developments, we have more powerful technologies at our disposal than ever before as individuals. This implies we have more power to change the world than ever before. We have more power to do the things that could not be done by earlier generations. In other words, we have the ability to shape the future. More than ever before, we can choose what we want as individuals.

The Impact of Automation and AI in Everyday Life

We cannot avoid the impact of automation and AI on our everyday life because consciously and unconsciously they are already there and we are benefiting from them. Some are constantly changing and improving over time. We can only track the present states of the art.

I am going to explore our new experiences in each of the following categories:

(a) Retail
(b) Personal and professional services
(c) Logistics and distribution
(d) Communication
(e) Cashless world
(f) Media
(g) Medicine

Retail

A lot of people, especially the young, are shopping online. Transactions are easy as you can use your credit card account and the retailer will deliver your purchase to your home. You are also given the liberty to change the item if you are not pleased with the purchase. Due to the use of big data, you can compare the prices of other retailers and make a sensible decision. This was not possible before. Another attraction is that there could be more choices as conventional shops tend to stock only a limited number of items and can run out. Online, and with get in time deliveries, there are infinitely more possibilities.

Online ordering of food from a restaurant is also getting popular. There is less hassle compared to getting into the restaurant and you can wait in comfort either at home or at your office.

Those of us who frequently go to supermarkets would experience faster checkouts because all the items have barcodes. You can also do the checkout yourselves. It was quite different as little as 5 years ago. In future where RFID chips are to be tacked to each item, the checkout process would be even more simple.

Another positive development in the future would be the use of AR (Augmented Reality) for retail. All you need to do is to point at an item and all the details can be read — expiry date and contents details. I believe NTUC Supermarket will have such facilities in the near future.

Due to the introduction of big data in retailing, your purchases will be tracked. The retailer will be able to analyse the types of goods you have purchased and will be able to analyse your purchasing habits and will introduce more to you through the internet or your handphone. This would be similar to the early days when you ordered books through Amazon. The next time you open up your account, you would see new recommendations — all primed to lure the dollars out of your wallet.

Another transformative force is the explosive growth of the fully automated self-service retail in vending machines and kiosks. Vending machines have progressed beyond dispensing merely snacks and drinks to almost any form of consumer goods.

Vending machines make it possible to dramatically reduce three of the most significant costs incurred in the retail business: real estate, labour and theft by customers and employees. The latest generation of machines are directly connected to the Internet and provide a continuous stream of sales and diagnostic data.

Personal and Professional Services

Another more sophisticated personal service assistance is "Robot waitress" at restaurants in Japan and South Korea. In Japan, robots are also used as aged-care workers but the use is still limited because you cannot automate caring which needs compassion and empathy.

Professional services on legal matters are facilitated by AI. Since legal decisions in law are often based on precedence, the Big Data software can assist in combing through past cases and identify the ones similar to the present case. That could help the lawyer to put up a good argument for the case.

Logistics and Distribution

A most visible benefit of the increased use of AI and automation is the delivery service. Greater efficiency, productivity and transparency in the order fulfilment process.

Most of us have the experience of making online purchases. We push a few buttons and the goods appear at our doorstep. In between, the retailers make use of a delivery service to get the goods out of its warehouse and into our homes.

It is a complex process to make use of the efficiency of a good distribution system. The system generates the best route of delivery as well as the order service it is likely to deliver a few items to different destinations. On a smaller scale, when we request a food delivery from a restaurant, Foodpanda or its equivalent will need to go through a similar analysis and process. A user-friendly software is available to help in the decision making.

On a larger scale and over longer distances, drones are used to give prompt delivery. This is found in larger countries like China. Shipping lines for cargo deliveries are from port to port. It is also aided by AI and automation software.

A charming development is the delivery-to-the-door robots at a number of 5-star hotels. The house guest makes an order through the house phone. Minutes later, a robot appears at the door with the item. This service has also been seen in some restaurants delivering the food ordered.

Communication

Communication has evolved rapidly from telephone, fax machines, the Internet, email to social media. In the years to come, communication will become very automated.

We shall find ourselves communicating directly with machines. A good example is Google Home. You get the machine to respond to all your questions and requests. For example, you can ask for all the news broadcasts as well as specific information. If you can find answers through the Google search, you would get the answers from the machine.

Another common device is the chatbot which is a computer-generated service assistant. It is often connected to an answering machine on the telephone of an organisation. Experts have estimated that AI powered chatbots had successfully interacted with customer services for 85% of the time. A real person will take over in the 15% when the chatbot fails.

Another use of AI assisted assistance is the use of an automatic language translation. Although the reliability in the present software is noted to be about 95%, it will improve overtime.

Use of automated communication not only saves costs as less manpower is required, but also saves time as most enquiries can be answered directly rather than being bounced around different parties.

Cashless World

The move towards a cashless world is a significant challenge and disruption to the financial sector. Telephone companies are participating in a big way through their Apps like Google, WeChat and Alibaba. It has even attracted companies like Grab to participate. The latest drive to set up Digital Banking will open up further to the players into the financial sector.

This development benefits many people especially in places where people live some distance away from bank branches. While these consumers may not have banking accounts, they have smartphones which allow them to do all the usual banking transactions.

This newly created on-demand economy, on the other hand, does bring about some new issues. One is taxation as it is much easier and more attractive for companies to operate in the black market. Another is money laundering as it is more difficult to track the source. Cyber security is another issue as the banking accounts have to be protected from hacking. Even with all those issues, the ease of operating in a cashless world will attract more and more people to participate.

Media

When we refer to the media 20 years ago, they were the newspaper, TV, radio and internet. All could be controlled by the government if it chose to do so.

Now with social media, information is free for all and transmitted to all instantly. Public opinions are often channelled through Facebook or Twitter.

Two visual media have appeared which become everyone's entertainment or information channels. The most powerful media is

YouTube which has more than 1.5 billion viewers every month. It is now the largest TV network in the world.

Next is Netflix which may one day replace broadcast TV. It is a source of TV entertainment for many people.

Job skills for the media would have to be changed to one where the programmer needs to understand Big Data to know the popularity of the programs and change if necessary.

Medicine

Medicine is an exciting field with the impact of AI and automation. The numerous advances in physical, biological and digital technologies give rise to development in diagnostic approaches and medical treatments.

The primary role of medical doctors is to give accurate medical diagnoses so that proper treatment can be given. AI software has been able to give accurate medical diagnoses, since the symptoms are inputted into the program. The doctor in attendance can then take over to begin treatment.

In surgery, automation is proving to be a useful aid. A large percentage of robots sold worldwide are designed for surgical purposes. It has been reported that more than one million Americans have undergone robotic surgery. The usefulness of robotic surgery is because they have been found to perform procedures with greater proficiency than humans. They are less invasive and lead to faster recovery for patients.

In time to come, development in wearable devices and implantable technologies will give rise to new diagnostic approaches and therapies. Digitising of patient records will be enhanced.

Education in the Era of Disruptive Technology

National University of Science & Technology, Oman (NUST)

NUST was formed in 2018 as a merger of existing colleges: College of Engineering, College of Medicine and College of Pharmacy. The establishment dates of the colleges were: Engineering 1996, Medicine 2001 and Pharmacy 2003. The total number of students at NUST was 4,000 with international students coming from 33 countries. The number of students in the respective colleges were: Engineering 3,000; Medicine 800; Pharmacy 200.

The Sultanate of Oman holds a strategically important position at the mouth of the Persian Gulf. It is a country with parliamentary absolute monarchy. The last Sultan Qaboos Bin Said ruled Oman from 1970 to January 2020. The land area of Oman is 3,095 hectares and a large part of it is desert. It has a population of 5 million with GDP(PPP) of US$47,000 per capita, has vast oil reserves (25th global ranking) and tourism is the next largest GDP generator.

Oman is a beautiful country at the mouth of the Persian Gulf. Here is a luxurious hotel at the coast.

In conjunction with the celebration of the formation of NUST, a National Lecture by a foreign expert was planned. I was invited for the inaugural lecture on 3 December 2018 and was given a free choice on the lecture topic. Being uncertain how receptive Oman is on a foreigner's views, I decided to give a lecture on a neutral topic and on an area of my personal expertise.

The lecture was "Education in the Era of Disruptive Technology." The topic was on an area I had been doing some research for the last few years. Since the rapidly changing technology would leave a significant impact on businesses and human lives, education needs to adapt to the changing needs of the students.

My lecture was well attended.

The Guest of Honor was the Deputy Prime Minister Sayyid Asaad, cousin of the Sultan. In the audience were 5 ministers, 8 under-secretaries and 12 ambassadors. There were also government officials,

I delivered the inaugural National Lecture (Dec 2018), which was attended by Guest of Honour DPM Sayyid Asaad (third from right), who is also the cousin of the current Sultan.

ambassadors, and business leaders. Seniors academics from the various universities also attended.

It was a grand affair.

After my lecture, I was presented a ceremonial kris which I was told is normally given to high-level guests such as foreign heads of state. The kris has a silver sheath and ivory handle.

After the lecture I was presented with a beautiful ceremonial Kris, with a silver sheath and an ivory handle.

As a consequence of giving the Inaugural National Lecture, I was invited to be a member of the Board of Trustees of NUST.

Oman is a country with beautiful scenery of its coastline as well as its desert. I was fascinated as I travelled in Oman over the sand dunes. With such natural resources, it is no wonder tourism is a big business in Oman.

"Education in the Era of Disruptive Technology"

Delivered at the First National Lecture 3 December 2018

"We are witnessing changes in ways of doing business, lifestyle and work. The changes will become more rapid with the prevalence of disruptive technology made possible by advances in computers, artificial intelligence and automation. This age of disruptive technology is sometimes called the Industry 4.0.

The first industrial revolution was attributed to the development of the steam engine which transformed the agrarian societies to the industrial economies. The initial benefactors were the mechanisation of the textile industries. Then came the improvements in transportation as applied to trains.

The second industrial revolution was attributed to the development of electricity and used by Michael Faraday for electric motor technology, convenient household use and telecommunication. Mechanisation in the factories were further enhanced.

Industry 3.0 brought computerisation and the information and communication technology (ICT). The internet was able to connect everyone all over the world for business or for social purposes.

Now, Industry 4.0 represents new ways in which technology becomes embedded within societies, driven by advances in computers, artificial intelligence and automation. Both work and lifestyles will undergo significant changes.

Some experts predict that it will make half the population jobless very quickly because robots or digital transformations will take over all the tasks. I have a different view. The steam engine, electricity and internet did not cause mass unemployment overnight. The changes took place gradually and gave people time to adjust. So too will be the case for the present generation.

We have to accept there will be great changes to the way we work and our lifestyles. That is inevitable. To meet the challenges, our students have to be equipped with skills by education and training.

We can no longer rely on the traditional education of just broadening the minds. Machines are getting more intelligent than humans and with their ability to store large amounts of information the machines are more knowledgeable than all of us. We need a different strategy of preparing our students so that they are better than the machines in other ways.

I believe we need an immediate as well as a long-term strategy in education to succeed in this era of disruptive technology.

For immediate implementation, we need to enhance the skills of our students to be able to make use of the power of the computers and machines rather than to compete with them. To do that, the core knowledge needed would be in IT and robotic engineering.

AI is related to the ability of machines for general learning and high-level reasoning which are cognitive functions associated only with humans. The key to the success of Artificial Intelligence is good computer programming on powerful computing machines. To develop our students towards those skills, we may need to review our high school curriculum. It may be necessary to include in all streams of studies a good grounding in mathematics and also sciences. This will not be a unique system as it has been practised in the French high school system since Napoleon days. The additional knowledge in mathematics and science will help the young students understand the abilities of the computers.

At a higher level, at the Polytechnics and universities, all streams of studies need to include modules on IT and data analytics. IT is the basis of the intelligent world. It comprises disciplines such as internet communication, Internet of Things (sensors), virtual reality and digital network for data communication.

It is difficult to avoid 'big data'. With the power of the Internet of Things, fast computers and enormous data storage capacity, data accumulates at a tremendous pace. From data, information is constructed. Collections of related data give insights into useful knowledge. If the knowledge is targeted, solutions to problems can be devised. So, data analytic has been applied in many areas such as business decisions and logistics, and it's a powerful scientific tool.

The tools are needed not just for technical people. Professionals such as doctors and lawyers will also need to rely on such expertise for their daily work in the future. Examples are in disease identification using AI, medical robotics, legal case search using AI, language translation of legal documents, and so on.

Other than equipping our students with technical skills, higher education needs to be more targeted towards professions and careers. Our universities and polytechnics should provide their students with more channels for internships which would cultivate students to be more market oriented. Our present national strategy of re-skill and up-skill, and skills future are in line with the need to adapt to the challenges of disruptive technology. We have to note that skill in this context does not refer to 'manual' skill only but should include 'knowledge' skill.

I like to re-emphasise the importance of acquiring skills for the future education of students and workers. We need to have the skill to make use of the power of the computers since we cannot compete with its intelligence. That is, we need problem solving skills on the best approach to get the outcomes.

Looking at Singapore's education system, I think Polytechnic education is best suited for problem solving. I venture to predict the best path for our students in the future is to go through a polytechnic training first before going to the university. The German education system has a similar philosophy.

For the long-term learning strategy, there are three aspects of formal education we need to emphasise. The first is the cultivation of creativity. Many people think that the human ability to come up with a new idea is better than machines. Ultimately, computers lack imagination or creativity to dream up a vision for the future. It lacks the emotional competence that a human has.

Thus, creativity will be the skill of the future. The key to staying ahead and participating in the creation of the future is our own creativity. We must embrace and develop our creativity, and then use technology creatively to solve problems of the world.

Our education system must nurture creativity of the students. This cannot be achieved by just learning facts to come up with new insights. I believe the best way to promote creativity is through open-ended projects. It is even better if projects are performed by a team. The interactions of members often generate new ideas and new approaches to problem solving.

The second requirement of higher education in this new age is to be able to mould the students to be good members of society. The students have to learn the art of giving and taking. This behaviour is superior to machines which have no moral agency and can act only through a prescribed program.

The education path is to develop Emotional Quotient (EQ) and not just IQ. EQ is the capability of individuals to recognize their own emotions and those of others and use such information to guide their thinking and behaviour.

Education modules should include studies on ethics, self-awareness, self-control, social skills and empathy. I am aware such skills are difficult to impart. This is often achieved through inspirational lectures or case studies and practical acts such as visiting old-folks' homes and homes for disadvantaged. Parents can also help at home by being role models.

The third requirement of a future education system is to include a strong emphasis on life-long learning. This is getting more urgent as time passes. New technologies and new concepts are introduced constantly. Knowledge will become obsolete very quickly and therefore, continuous learning is needed to make oneself relevant. To enable learners to keep up, using E-learning is a good platform.

All public and private bodies need to have E-learning facilities as well as the support services to make such learning methods easy to use.

To sum up, education in the era of disruptive technology needs to incorporate learnings on skills to take advantage of the intelligence of machines as well as teachings in creativity, development in EQ and an effective system of continuous education. These are means to make humans better than the machines. Until we have an effective strategy on education, we would be left behind by Industry 4.0.

A concluding note is that even in a world full of powerful hardware, software and networks, people remain valuable because of their judgement and reasoning beyond calculations and data. Therefore, it is important that they have the best education in knowledge and skill."

I spoke on technology and education which saw a flurry of questions, including from the Minister of Higher Education.

I took the opportunity to visit Oman, including tasting some fine dates and riding a camel (page 160).

Chapter 9

Significant Personalities

Mr Lee Kuan Yew (1988–1991)

For many years, I was fortunate to have a monthly lunch with Mr Lee Kuan Yew. They were held in a small room next to his office at the Istana. Sometimes, Prof Lim Pin joined me. The dining room had a small table seating for 3 persons at any one time. On one side was a number of chopsticks in a chopstick stand. Gloves were placed on top of each of the chopsticks which would be used by LKY before slicing the juicy fruits.

They were for the protection of his delicate fingers.

A typical lunch would be very light, consisting of two slices of potato chips, a watery soup and a plate having a tomato, some green vegetables and a thin slice of meat, usually chicken. A cup of Chinese tea would be included. The lunch would finish with a basket full of fruits such as honey jackfruit, pomelo, pear and orange. His philosophy of a suitable lunch was one light enough to allow a pang of hunger before dinner.

The main part of the lunch was the conversation over many topics. I believe he was interested in our views in issues raised by him. The topics covered a wide range and surprisingly very little on education knowing that I was the President of NTU. Of course, most of them were in Singapore but because he was travelling widely during that period, he showed us his views on these countries.

I value the conversations because to me they were truly tutorial lessons on worldwide issues and politics. He was a visionary and many of his viewpoints are still valid today.

I was particularly keen to hear his views on the future of Singapore. He thought that our needs were small. Even if we faced world disasters, Singapore would be alright if all of us worked hard. He was worried about the younger generation who might be more comfort-seeking rather than hard working.

He was glad the men had to go through NS which would make them realistic.

Since I was from the University, I did ask him a few questions on university education. One was to plan university intake according to economic needs. His view was that planning on economic need was an ideal solution.

PM Lee Kuan Yew took a keen interest in NTU. (In white jacket visiting NTU, 5 Apr 1990).

Walking along the North Spine. I would have regular lunches with PM Lee while I was the President of NTU.

Every parent would want their children to be a graduate. So, the universities should teach the student well no matter which course he took. If they are taught well, they can rise to a country's needs easily. This, he added, was the US's experience.

I was interested in his views on the role of government. He listed 3 and elaborated on them. The first was economy, the second was military and security — to protect Singapore from external pressure, we need a strong military force. That was the basis of our NS. For internal security, his views were that the only way was through racial harmony and good grassroots initiatives. The third was media. His concern was to have truthful reporting and proper explanation and communication to the people about government policies. (Of course, nowadays it is not going to be that easy with many messages from different sources going out social media.)

As a lunch host, he was as good as any I have seen. As a lunch companion, he was definitely different from his rather intimidating public persona. I had never seen him losing his temper over the lunch although sometimes I did say rather stupid things. Overall, I found him like my colleagues but infinitely wiser. I did learn a lot from him, and I missed him when he stopped lunching with me.

Dr Goh Keng Swee

I met Dr Goh the first time in 1975 when he was the Defence Minister. He got me and Prof Louis Chen of the Mathematics Department at NUS to work on a MINDEF project. It was a study on the precise delay required in the explosion of a mortar bomb.

The issue was that when a mortar bomb was launched and hit a tree, it took a small amount of time to hit the ground. Ideally, we'd want the bomb to explode *only* when it hits the ground, as the explosion would be more effective at ground level. To work this out, we carried out a simulation, and then worked on a statistical study on the characteristics of rubber trees.

Then the timing of the fall could be worked out, and that became the delay of the fuse needed. At the completion of the project, we met Dr Goh to discuss findings. He was very pleased with our study.

The next occasion was in 1979 when he was the DPM as well as Minister of Education. I was then the Dean of Engineering at the University of Singapore. He called me to his office in Kay Siang Road to seek my opinion of AIT (Asian Institute of Technology) in Thailand. He was invited to be on the Advisory Board and asked me for some facts about the University. He was reluctant to accept the appointment but told me he had no choice because the Chairman was a former Prime Minister of Thailand.

The next encounter with Dr Goh was not so pleasant. This was in 1989 when he was the Chairman of Tote Board. Dr Tony Tan agreed to give the MIT, Sloan School US$10 million for giving assistance to NTU's School of Accountancy.

I had to approach Dr Richard Hu, Minister of Finance for the money. In turn, Dr Hu instructed the Tote Board to come up with the money. Dr Goh was not happy that NTU gave away so much money for what he felt was little assistance from MIT. He phoned me and gave me a scolding, telling me that I did not appreciate the value of money.

In the end, the Tote Board did give NTU the US$10 million.

In 1994 I met Dr Goh under more friendly circumstances.

He was the Advisor to the UOB Group and Mr Wee Cho Yaw invited him and I for lunch at his UOB office. Dr Goh asked me about the progress of the development of NTU. I told him NTU was doing well but would appreciate some financial support for its laboratory.

A month later, he instructed the Tote Board to provide NTU $10 million to equip its manufacturing laboratory. Dr Goh thought that NTU was producing useful graduates as compared to Nanyang University.

In 1995, I met Dr Goh for the last time when he was the Chairman of the Institute of East Asia Philosophies. The Institute was then in NUS and was running out of space. It was at that time, NTU was looking into the redevelopment of the old administrative building (the 2-storey building which had a distinctively Chinese architecture) to house the institute. The redevelopment had to be abandoned because there was not enough land. Therefore, the Institute remained in NUS.

Overall, my impression of Dr Goh was he was a sharp person who could see through the complexities of any issue. He could be gracious to people but had a quick temper. You could enter the room with a cheerful Dr Goh and leave after being chased out. All those who worked closely with him would have had this experience.

Dr Goh got on well with Sim Kee Boon and Wee Cho Yaw.

Dr Toh Chin Chye

I met Dr Toh after becoming one of the pioneer 13 staff who started the Faculty of Engineering of the University of Singapore. He was the Vice-Chancellor. He somehow identified me as someone worth cultivating.

Being an engineer by training, he thought it would be desirable for me to be involved in the Institution of Engineers Singapore so that the Faculty of Engineering could have an influence in the local institution of engineers. I made an application to be a member but was rejected because I had no practising experience.

Dr Toh Chin Chye visiting the Engineering faculty, University of Singapore (1978)

This was a departure from usual practice as the UK Institution did count research experience as equivalent to practising experience. When Dr Toh heard this, he intervened and I was later admitted as a member. Subsequently I was active in the Institution and succeeded in holding some key positions in 1972.

Dr Toh was kind enough to allow me to return to Cambridge for a three-month research contract working under my previous PhD supervisor. I did manage to learn new things in particular how to use towing tank equipment to study the movement of a sea craft in water.

On my return to Singapore, Dr Toh thought my new knowledge would be useful for starting an Ocean Engineering Centre in Singapore. Singapore being a seaport and had many repair and shipbuilding yards would be in a position to start a world-class Ocean Engineering Centre.

I was sent to Japan and Hawaii to look at equivalent centres. On return, a proposal was made and a potential site in Jurong was chosen. However, by that time, Dr Toh was no longer the Minister of Science and Technology as the Ministry was eliminated, so the project was shelved.

Dr Toh, as a scientist, was logical in his approach to problems. However, he was stubborn and not prepared to compromise. I understood that he had many arguments with Lee Kuan Yew. On his retirement from government, I still kept up with him when he was an Advisor to Keppel's Data Centre, and when asked, provided some inputs to his work.

Tan Chin Tuan

I first came to meet Mr Tan Chin Tuan when he summoned me to his office in the OCBC building in 1986. I think he was interested in finding out more about me after I joined the Wearnes Brothers Board in 1985.

At that time, Wearnes was a company within the OCBC group of which he was the Honorary President. We chatted on general topics, and he found out about my background, my training and my work in NTI. Mr IF Tang, chairman of Wearnes told me Chin Tuan was curious what sort of person I was — being a Director of Keppel as well as President of NTI.

We did not meet one-on-one for some time after that, although, there were some occasions and functions to say hello to each other. It was from 1996 that we began a series of meetings and lunches.

This was held at his home at Cairnhill Road. He would invite me for a meeting about an hour before lunch. There were always three of us, Chin Tuan, myself and Gek Khim, his granddaughter. She was there to take down notes on what we discussed. We talked on topics like geopolitics, world economy, currency fluctuations, Singapore economy and Singapore companies.

He also liked to know what was going on in our education system, schools, polys and in particular NTI. I sensed that he was a strong believer in good education and therefore created many scholarships in schools, polys and universities as part of his philanthropy.

Because I was president of NTI, he was prepared to help me in developing NTI engineering professors. He donated $10 million to NTI/NTU mainly for the Tan Chin Tuan Exchange Fellowship for Engineering as well as the Tan Chin Tuan Professorship in Engineering. The Exchange Fellowship was to enable young Singapore professors to have exposure to overseas universities in research.

The scheme also provided for eminent foreign professors to come to NTU to help in our research. Both initiatives are still very active and successful up to today. I am responsible for monitoring the progress on behalf of the Foundation, in whose name the donation was made.

Although he was not a person interested in the arts, as Chairman of SSO I approached him for a donation and he was quick to respond. He even informed Gek Khim that whenever I approached the Tan Chin Tuan Foundation for a donation to SSO, she should consent. This is the case until now.

The pre-lunch meetings took place at first on a monthly basis and then progressed to a weekly basis. At the lunch about 10 other persons joined in. Very frequently, Mrs Mary Tan (wife of Tony Tan) was present, as Tony Tan is the nephew of Tan Chin Tuan.

It was at this time that he invited me to be a member of the Foundation. There were five members of the Council of the Foundation. The Chairman is his daughter Dr Chew Keng Lian and three others are his grandchildren. I am the only non-family member. The Council meets quarterly and decides on donations to be given. The two largest groups to receive donations are Education and Community. Typically, the yearly donations come up to about $3 million.

It was also at this time in 1998, he wanted me to consider being the CEO of his personal investment fund. At that time the size of his investment fund was about $800 million. His goal was to exceed 1 billion.

I told him I still had a lot of work at the NTU and could not accept his proposal. Nevertheless, on my retirement from NTU at the end of 2002, I was offered a position as an advisor of the investment group. It was a great experience for me to be exposed to a large investment fund of about $1.7 billion which had grown from $800 million in 1998. To be effective, we had to meet twice a week.

Tan Chin Tuan was a great banker and renowned philanthropist. In his lifetime, he made significant impact in Singapore history by participating in the colonial government. He passed on in 2005, and I lost a great mentor.

Mr Wee Cho Yaw

Mr Wee Cho Yaw is a renowned banker associated with the building up of United Overseas Bank. Many consider him as one of the top bankers in the history of Singapore. Other than being known as the Chairman of various Chinese Associations, he is also remembered as the last Chairman of the Nanyang University Council.

I met him initially as the President of NTI. He hosted a number of lunches with the Nanyang University Alumni Association Officials in the 1980s which included me. The purpose was to discuss ways to preserve the legacy of Nanyang University. It showed the fate of Nanyang University still hold dear in his heart. Through the discussion with Mr Wee, NTI arranged some of the activities of NTI to include past graduates of Nanyang University.

It was in 2001, that Mr Wee invited me to be a Director of UOB. As a result of this appointment, I worked very closely with him for almost 15 years being a member or Chairman of every subcommittee in UOB. Due to the close working relationship, I could observe Mr Wee as a banker. He has a deep knowledge of the banking industry, having been in the banks for his whole life.

He has a good intuitive sense about the prospects and risks of various aspects of banking. I presume it was developed through his vast experience and from meeting and talking to a great number of people in the industry. What impressed me most was when I met MAS officers due to my position as the Chairman of Audit Committee — these officers often asked me what his views were — it showed me that MAS has a deep respect for him.

Mr Wee also invited me to be on the board of UOB (China) since 2006 which gives me opportunity to know China.

Michael Fam

Michael Fam's reputation preceded him. Among many of his roles was the Chairman of HDB which was a high profile one. But it was in his

Michael Fam (left) was a great mentor to me, when he was the Chairman of NTU (photo taken in 2005).

role as Chairperson of NTI Council and later NTU Council that I had the opportunity to work closely with him.

When NTI was established in 1981, I had to look for a Chairman to lead the Council. Dr Tony Tan, the Minister-in-charge gave me two choices: Michael Fam or Dr Richard Hu, who at that time was Chairman of Shell and a chemical engineer by training.

I chose Michael because he was then Chairman of HDB and a civil engineer. Being a new university, I expected to deal with a lot of physical infrastructure development. Michael's expertise would be useful to NTI. Dr Tony Tan agreed. So, he was Chairman of NTI Council for 10 years from 1981 and NTU Council for 2 years from 1991.

Michael did more than help me develop NTU — he was also a great leadership trainer — I learnt many business strategies from him. One of the important lessons he gave me was to have a good understanding on the roles of the Board and Management of an organisation. The Board is to deal with policies to guide Management and Management has to implement the mission of the organisation.

The two cannot be in conflict.

If there were a conflict, you either sacked the Board or the Management. It was under this principle that he governed NTU. He left most of the management of NTU to me and all I had to do was to keep him up to date.

My discussions with him often lead to new business strategies because both of us were active on corporate boards. He was executive Chairman of F&N, SIA and directors of OCBC Bank and SPH. I was director of Keppel Corporation, Wearnes and Chairman of Natsteel. These were many common interests in the corporate world. We often discussed how to evaluate the worth of a company and ensure a satisfactory return for shareholders. I was glad that I could learn from an experienced corporate leader.

I have to acknowledge two useful contributions for NTU from Michael Fam as its Chairman. First was the ability to negotiate with the Phase I construction contractor Kumagai Gumi. Due to an unforeseen soil condition, there was some difficulty in the construction of some buildings.

The contractor demanded a somewhat exaggerated amount for payment for the additional works. Michael, having experience of dealing with civil engineering contractors, very quickly settled the negotiation for a reasonable sum.

The second was on accommodation of students in the campus. In 1988, NTU succeeded in moving the School of Accountancy from NUS. One of the concessions was to provide hostel accommodation for those students affected. There were a number of unoccupied HDB blocks outside the campus. To use the HDB blocks at a short notice could only be made possible by Michael Fam who was Chairman of HDB.

I continued to meet up with him socially until he passed on in 2014. He was in poor health for a number of years. Despite being confined to a wheelchair, I am very grateful he took the trouble to be present at the launch of my book "The NTU Story" in 2014.

Mr Tang I-Fang

IF Tang first came to Singapore from China, leading a UNDP program team to give consultation to Singapore on how to industrialise its economy. This was in the early 60s.

He settled in Singapore and became EDB Chairman from 1968 to 1972. It was in the early 70s before I had a chance to meet up with IF. He invited me to attend a few meetings at the EDB where panel discussions were conducted on various manufacturing industries likely to be attracted to Singapore.

IF Tang, Chairman, WBL, (pictured here at Wearnes Dinner & Dance) first came to Singapore in the 1960s as part of a UNDP mission.

I think he was impressed with my knowledge of Engineering and invited me to be a director of WBL Corporation in 1985.

At that time, the main business of WBL was in automobile agency. It had famous brands like Rolls-Royce, Jaguar, Volvo and Ford. It had plans to diversify into high tech which I could contribute since the other existing Directors were not technologically trained.

The first entry to technology was the production of simple-powered computer with the brand name of Wearnes Technology.

IF being an engineer by training was very proud of its first success in IT. With the advances in electronics, WBL eventually went with the semi-conductor businesses. The credit should go to IF for his visionary insight into high technology.

IF was also Chairman of JTC (Jurong Town Corporation) from 1979 to 1986. In 1985, he brought me into JTC as a Director because he felt my role as President of NTI was helpful for the corporation. NTI professors could interact with the JTC management for mutual benefits. I remained in the JTC board until 1996.

Mr Tang decided to retire from the WBL Board in 2006. He wanted me to succeed him as the Chairman and for that he had to consult Mr Lee Seng Wee who represented the main shareholder. Mr Lee was reluctant to have me as the Chairman as WBL was seen to be in the OCBC Group. I was at that time a Director of UOB and UOB was seen as a rival bank to OCBC.

It would be an embarrassment to the OCBC Group if I succeeded Mr IF Tang. Therefore, the idea was dropped.

I saw very little of IF after he left WBL. It was sad that he passed on in 2013. He was a great friend and mentor.

The Riady Family

This relates to Stephen Riady's father Mochtar and brother James. I met Mochtar for the first time at a lunch hosted by Wee Cho Yaw. I talked to him briefly and even though it was a brief encounter I sized him as an old man with great authority. He was confident and at the same time gracious. The family is well-known in Indonesia as the Lippo Group.

My first encounter with Stephen was not a pleasant one. Due to the hostile takeover by his company on Robinson, the department store, I gave some uncomplimentary remarks to the press which was published in the Straits Times the next day. For that, he proposed to sue me for defamation but he did not take action subsequently. Fate decided that we became good friends eventually.

The family owns a private university, Universitas Pelita Harapan, in Indonesia which has been ranked as the best private university in Indonesia. I was invited to be an Advisor to the university which I accepted, and still remain till today. When NSW of Australia abandoned the Changi campus, UPH wanted to build a Singaporean campus in its place. In fact, I had begun worked with its architect on the design.

Unfortunately, the campus was already embarked for the SUTD and the project did not take place. I continue to engage Steven and he accepted my invitation to be on the SSO Council. He has been gracious in supporting SSO.

I did fly to Mochtar's home for a visit in Karawaci by helicopter from Jakarta Airport. As expected, it is a huge house in a large garden. There I met James, his son. Subsequently I met him again when he dropped into SIM. He is a person easy to make friends with.

Mr SR Nathan

I met Mr Nathan for the first time in 1981 when he was the executive chairman of the Straits Times. He was introduced to me by Mr Michael Fam, the Chairman of NTI. We met SR in his office in ST.

The purpose was to get ST to provide publicity to the new institute NTI. He was very cooperative and immediately assigned Gretchen Liu to work out the task. We were very thankful for ST help as the public got to know NTI well quickly and received acceptance as an IHL.

We met a few times when he was on the High Commission for Malaysia but it was only in 1996, I really got to know SR well. NTU was looking to establish its Institute of Defence and Strategic studies in 1996 and we were looking for a Director. At that time, I heard from MFA he was going to retire as the Ambassador to the United States. I consulted Dr Tony Tan and Dr Tan was positive about him. Dr Tan advised me to convince him to take the position as there would be other offers for such an eminent person. I flew to Washington and discussed the possibility with him. He was in fact very modest and said that he did not have a degree and could not direct a Think Tank which had to do in depth studies of strategic issues.

I told him he needed no qualifications other than his role as Ambassador to a most powerful country. Furthermore, his broad network would be useful to the Institute. Finally, he agreed to take up the position when he returned. I offered him a 3-year contract renewable on expiry.

He proved to be an exceptional Director of the IDSS which enabled the Institute to be known worldwide.

I had in fact renewed his contract for another 3 years, but during the first year of the second contract, he came up to me and told me he had been asked to stand as a candidate to succeed Ong Teng Cheong as President.

At the same time, he asked me to be his Proposer and if there were a contest, I had to be his Campaign Manager. Of course, I was happy to be his Proposer. When his first term was up, he again requested me to be his Proposer for the second term. There was no contest for both terms.

Although he was from a minority race, he proved to be a popular President. He could get along with people of all races and all walks of life. One of my friends who frequently walks along the East Coast had met him a number of times. He would not fail to stop and had a chat with him, although he did not know who he was.

During the period as Singapore's President, I met up with him frequently either officially or privately for lunches. He nominated me to be a Member of the Presidential Council from 2004 to 2010. The Council and President made frequent visits to government institutions like GIC, MAS and Temasek Holdings. He was quick to identify the important points in the role of the President as custodian of our national reserves.

When SR retired in 2011, he took the position of Advisor to SMU and had an office there. I kept up our contacts either by going to his office or him coming to see me in my office in SIM. It was sad he passed on in 2016. He was a friend, a colleague and a mentor.

Sim Kee Boon

Before I met him, I knew of him as a high-profile civil servant. He was permanent secretary of many ministries. It was when he became Head of Civil Service (1979–1984) that I had the opportunity of meeting him as President of NTI. He was able to communicate very clearly the needs of Civil Service and how each ministry or statutory board can contribute.

It was in 1984 when he retired from the Civil Service and became Chairman of Keppel Corporation that I got to know him well. Being on the Keppel Board since 1982 and Chairman of the Audit Committee, it was thought I would be the appropriate person to bring him up to date. I spent one full day going through the businesses of the various Business Units and the work of the different departments in the central management. I also briefed him on the business risk of the Group as it was then one of the worst periods of the Singapore economy. He was quick to grasp the situation due to his training in economics and working in the Ministry of Finance.

In the course of the next few months after taking office in Keppel, he was quick in rectifying the different financial conditions of Keppel. He sold the whole fleet of ships and was involved in the proposed sale of the Ocean Tower.

Fortunately, we could not find a suitable buyer of the Ocean Tower because within the next 6 months, the property market turned up. Instead of selling, he redeveloped it by having a twin tower.

He diversified Keppel's business to real estate (Keppel Land), banking (Keppel Bank) and others like telecommunication and data center.

My contribution was through my knowledge in technology. I urged him to start M1, the telco — in fact, I was awarded the first two mobile numbers from M1, which my daughter and I kept till today. Kee Boon also sought my advice regarding the radar system of Changi Airport since he was also Chairman of CAAS. He was always quick in adopting new ideas for the company.

I was a member of the Council of Presidential Advisors in 2004 of which he was the Chairman until 2005. We had good working relationship since 1984. He passed on in 2007.

George Bogaars

My first encounter with George Bogaars was not a pleasant one. This was in 1981 when he was the Permanent Secretory of Ministry of Finance. I had to see him regarding funding for NTI. He was not sympathetic to my desire to put NTI on par with NUS. My only course of action was to consult Dr Tony Tan on how to resolve the problem. Dr Tan made an approach to Mr Hon Sui Sen, then the

A rare photo of George Bogaars and I, digging for tasty morsels at a Keppel Function (1984).

Finance Minister. Mr Hon's view was that he had to go along with his Permanent Secretary. Finally, Dr Tan had to appeal to PM who eventually overruled the Ministry of Finance.

In 1982, I was appointed a Director of Keppel Corp where the Chairman was George Bogaars. As a Chairman of a company, he wanted directors to help him grow the company. I did help him in Keppel by being a part of his team. We worked well together.

As a Chairman of a company, he had the foresight to look at proper governance of a complex company. Keppel was not only doing ship repair but was into building of oil rigs, ship ownership and property development. We were not sure our finance department was up to the mark to deal with the complexity.

He retired in 1985 and was succeeded by Sim Kee Boon.

Lee Seng Tee

Mr Lee Seng Tee is one of the famous three Lee brothers, sons of the late Mr Lee Kong Chian. They are the founders of the Lee Foundation. Seng Tee is the only surviving brother.

Seng Tee is a learned man. Surrounding his open-space office are shelves full of books. They are on a wide range of topics, some in Chinese and mostly in English. He is known to have donated with his own money to world-renowned universities like Cambridge and Wharton. I managed to get him to donate towards NTU too.

Despite his wealth and status, Seng Tee is very friendly person. When he was younger and fitter, he used to host Wednesday males-only lunches in his office. There were weeks when I was busy on Wednesdays, he very kindly hosted me on other weekdays. At the lunches, we could chat on a wide range of topics.

Of course, he was interested in NTU and I suspect that this was the main reason he wanted to see me. Now that I am no longer in NTU, we tend to talk on Singapore politics and worldwide events. As Chairman of Lee Foundation, he has been quite generous towards my appeal for donations to worthy causes.

Chapter 10

Reflections

I am over 80 years old. I have gone through a wide range of experiences from changes in governments, impact of technology on our daily lives, change in entertainment media to various life-threatening crises.

I hope my account would be of interest to the younger readers, growing up in a new and different world, but knowing how we got here and why.

My first conscious memory of my life was when I was 3 years old in 1942. It was during the Japanese occupation. I remember when I had to hide in our home-made air bombing shelter when the public alarm sounded. I was too young to be fearful of the danger.

I remember we grew our food on a small patch of land in front of our house — tapioca and kang kong. Although they were not our daily food, they grew easily and helped in giving variety to our diet. For our daily ration of rice and vegetables, I had to queue at a public stand. My parents had to give private tuition in Chinese and English for kids in the neighbourhood to get some income for our other necessities.

There was great hardship but we got by somehow.

The British came back in 1945 and we were ruled by successive Governors from Britain. The people felt that the British were not really interested in the welfare of Singapore. That brought about the founding of the People Action Party in 1954 by Lee Kuan Yew.

Although some local participation in the governing of Singapore was made in 1955 by appointing David Marshall as the Chief Minister, it was not satisfactory. He was succeeded by Lim Yew Hock from 1956 to 1959. There were a number of scandals by Ministers in his office. Overall, there were a lot of corruptions in the government much to the unhappiness of the people. In 1959, there was the first full scale election to elect a full internal self-government except for internal security matters.

As expected, Lee Kuan Yew's PAP won the election and he became the Prime Minister. His first task was the eradication of corruption in the civil service which brought about efficiency in providing service to people.

The next big event was when Singapore became part of Malaysia in 1963. During the decade before Independence in 1965, there was great unrest in Singapore. Most well-known were the Hock Lee Bus strike, Chinese high schools boycott of classes and racial riots. That was my first experience of a curfew.

Singapore became independent in 1965, through a decision of separation by the then Prime Minister of Malaysia Tengku Abdul Rahman. The main reason was the disagreement on the rights of Malays in Malaysia. Eventually, Singapore being led by Lee Kuan Yew and his able team made Singapore a first-world country with our GDP growing very rapidly.

Technology and Impact on our Life

Using technology for work and information started in the early 1960s by the invention of hand-held electronic calculators. The calculators could automate a lot of calculations superseding the slide rules. The calculators could be programmed for all sorts of functions and data. The most famous brand was the Hewlett Packard electronic calculators.

Use of computers for scientific work and businesses were started in the sixties as well. When I was at Cambridge working on my PhD work, I had to do a lot of calculations and was glad I could use Cambridge's Titan, which was the most powerful computer in the world.

However, the memory was limited only to 50K words which is tiny even compared to our handphone. To overcome the limitation, I had to be innovative in my programming so that I could get the results after a short computation. Because of my exposure to computer usage, I was appointed the Director of the Computer Centre of the University of Singapore on my return in 1969. I left after one year because I was more interested in engineering work.

Home computing was very limited. I recall in 1976 I bought a Radio Shack TRS 80 desktop computer with an 8K memory. The external drive was a tape recorder. As expected, the computing power was very limited. I was glad I bought the desktop computer, as my son was able to learn about computers when he was in Primary 6.

With the world transiting to the modern internet in the early 90s, desktop computers and notebooks were getting popular. They are so powerful nowadays that there is very little need for computer mainframes. Users just log in without the need to worry about the network which is the work of specialists.

Telephone communication went through huge changes in the last 50 years in Singapore. I still remember the first fixed line telephone in our home in 1959. Telephone charges were quite prohibitive in those days. My parents had to ration our usage.

Then came handphones. I bought my first handphone in 1985, the Motorola "brick" handphone. It was called that because holding it was like holding a brick of 800 grams (modern handphones are less than 200 grams).

The next famous handphone was the Nokia "banana" 8110. They were 2G to 3G phones. Over the years, I had to keep up with the latest available in the market. Earlier, they were predominantly European or American. Then came the South Korean phones and now the Chinese phones like Huawei, Oppo, Vivo and Xiaomi. Now we are talking about 5G phones, although the telephone companies are not really ready yet.

Nowadays, it is difficult to find a person without a handphone. They are mainly used for email, sms, Facebook, Twitter, WhatsApp and WeChat. Voice communication using the handphone is becoming rare. This is the age of the new social media with instant information exchange.

Entertainment and Media

My experience began with watching movies in the cinemas in the 1950s. Near my school (RI) were the Capitol Theatre and Odeon Theatre. Tickets were priced from $1 to $3 depending on the class of seats. As a student, I could only afford the front seats of $1 which could be quite uncomfortable after watching for more than an hour.

We tend to go to the cinemas after the morning school provided there was no extra-curricular activity for that afternoon. Family outings to the cinema would be during the weekends or at night. The treat we always looked forward to was after the show, our parents would invariably bring us to a "milkshake bar" which served milkshakes, ice cream and cake.

One habit of the patrons to the cinema shows, no longer practised now, was that when the hero of the show performed the requisite acts of valour (customarily rescue of the damsel in distress), all the patrons would erupt into loud clapping to show great satisfaction. It must seem strange to present day youngsters.

Television in Singapore only started with black and white broadcast in 1963. In the beginning, the sessions were only in the evenings and mainly shown in English. The government was using this medium to inform the public.

Colour television started only in 1974. They served as entertainment for the whole family, initially only one English-language channel, which then expanded to other languages eventually.

There were not many choices for the programmes until the introduction of paid-TV channels. Another visual entertainment source was the use of VHS video tapes of movies or music performances. They were very popular at the time but considered obsolete now. Another alternative is the playing of DVDs which are still popular now.

Another source of entertainment popular at the time to the older Chinese was Rediffusion. It was a voice broadcast and a very popular program was storytelling in Cantonese by a well-known story teller from Hong Kong.

National Crises

I shall give some details of 6 significant national crises which had affected me in one way or the other.

(1) *Oil Crisis 1973 and 1979*

OPEC proclaimed an oil embargo targeting nations perceived as supporting Israel. Oil prices rose nearly 400% overnight which caused the stock market to crash. Inflation rose by 20% and the world was in recession. As a consequence, Singapore GDP growth slowed down to 5% from the previous year of 12.7%.

I was involved in the designing of the air-conditioning system of the new campus at Kent Ridge for University of Singapore. Dr Toh Chin Chye, the Vice-Chancellor of University of Singapore, deemed that we should be energy thrifty in line with the national message. He recommended there should not be air-conditioning for the new campus. We had to convince him certain areas needed air-conditioning not because of personal comfort but for maintenance of sensitive equipment. That was in 1974, the midst of the crisis. Fortunately, the restriction was released later and more buildings like the lecture theatres were air-conditioned.

Another personal experience with the energy crisis was the designing of the MRT system in the early 80s. As an air-conditioning engineer I was responsible for the design of air-conditioning systems for the MRT underground stations.

Again, due to energy conservation, the Ministry of Finance ruled that there should be no air-conditioning in all underground stations. I had to argue that it would be unbearable for the commuters which would make the government unpopular.

Finally, we compromised — there would be air-conditioning but set at a higher temperature. Furthermore, the underground stations would have sliding doors which could prevent cold air leaking into the tunnel thereby saving energy. So, the sliding doors were not really for passenger safety.

(2) *Recession 1985*

In 1985, Singapore experienced its first recession since independence. Prior to 1985, its growth rate was 8.5%. The recession was caused partly by the high wage policy of the 70s and its overvalued currency that snipped away its competitiveness. The GDP growth was negative

1.7% and the government took quick action to cut CPF contribution from 25% to 10% and recommended a wage restraint for 2 years.

Personally, I experienced my first commercial shock because 1985 was my first year as Chairman of a public company, Singapore Automotive Engineering. The painful part was I had to retrench about 600 workers in the company. Later, I discovered one of them was the father of a close friend, who suffered a heart attack because of the retrenchment.

The unemployment rate in 1985 was 4.1% for Singapore, much higher than the usual rate of 2.9%.

(3) *Asian Financial Crisis, 1997 to 1998*

The crisis started in Thailand with the financial collapse of the Thai baht. Capital flight ensured an almost immediate international chain reaction. Singapore could not escape the effects and growth fell from the normal 8% to 1.5% in 1998. (In fact, negative growth for the second half of 1998). Unemployment rose to 3.2%, although marginally. Property values fell and so was the stock market.

The Singapore government implemented two key relief measures — lower business cost and help for the individuals and households for 2 years. Our economy recovered by 1999 with 7.2% growth.

At the time, I was the President of NTU. Due to our prudent use of funds, we had over the years accumulated a reserve of $80 million. In 1998 at the height of the crisis, the Ministry of Education issued an instruction for us to return the surplus. However, we managed to convince the Ministry we had already earmarked the fund for our own development.

$50 million was used to build the Nanyang Executive Centre, $10 million for the Wi-Fi network for the whole campus, $10 million for the Nanyang Technopreneurship Centre and $10 million to fund the Nanyang Satellite Centre. The reserve fund was put to good use, much appreciated by the students and staff.

(4) *SARS February 2003, 238 cases, 33 deaths*

SARS was a serious epidemic but thankfully contained within 4 months. Victims had very high fever and therefore were easy to identify. The Singapore government was quick in imposing some stringent measures like quarantine, school closures, contact tracing, etc.

Businesses were affected by people choosing to stay at home. My uncle-in-law had visited Hong Kong at the peak of the epidemic and returned with a high fever. He chose to stay in a hotel away from his home. He subsequently recovered.

(5) *H1N1 2009. Influenza Aviara, also known as the Spanish flu. 194 cases, 18 deaths*

Singapore in April raised the DORSCON from green to orange and implemented contact tracing, treatment and vaccination. Infected persons typically had very high fever. My son went for a short visit to Zurich in April and returned with a high fever and was suspected of having the H1N1 flu. His wife and two children caught the flu subsequently. All of them recovered within a month. They just stayed at home.

Overall, life was normal for most people.

(6) *Covid-19*

Covid-19 has become a fearful worldwide epidemic because of its ease of transmission. Almost every country is affected in one way or the other. At this point of writing (31 May 2020), the worldwide infection has reached more than 6 million cases, with deaths of 104,000 in the United States, 38,000 in UK and 33,000 in Italy.

Singapore has 23 deaths. It is one of the worst epidemic in the history of the world.

Symptoms included fever, coughs and difficulty in breathing. Asymptomatic infection is also possible which poses some difficulties on whom to isolate. Earlier Singapore had raised its DORS alert to Orange and had proceeded to impose a lockdown called a Circuit Breaker (CB) from 7 April to 1 June.

During the period of CB, most businesses ceased operation and office work is to be performed online from home. Schools are closed and students are on home-based learning.

There will be a slow relaxation of restrictions from 2 June and this rate would depend on whether we can control community infection. So far, infections had been confined to dormitories of foreign workers.

Covid-19 started in China in January 2020 and by now the whole world has been affected. This novel virus has taught us to be careful with our relationship with the outside world and even to each other on social distancing.

No one knows exactly the overall effect but there will be inevitable changes in healthcare, the economy and our lifestyles.

******

My "Reflection" is to share with readers the experiences I had for 80 years. Your future will be very different from mine. I am sure it will be a brighter one.

Cham Tao Soon

May 31, 2020

I received the Public Administration, Gold Medal, from President Wee Kim Wee in 1986.

Some 17 years later, I received the Distinguished Service Order, National Day Awards (2003) from President SR Nathan.

I was chuffed to receive an Honorary Degree from the University of Loughborough in 1996. It is the most well-known University in the UK for sports, producing many Olympians. It had a special relationship with NIE for helping with our School of Physical Education.

A few years after I stepped down, NTU awarded me the title of President (Emeritus) (2007).

Presenting Mr Lee Kuan Yew with the inaugural RI Gryphon Award (2011).

Convocations are among my favourite times of the year. Here with President Ong Teng Cheong and Education Minister Lee Yock Suan.

In 1987, David Marshall (then Singapore's ambassador to France) received an Honorary Doctor of Laws degree from NUS. I had the chance to work with him when the PSC began awarding scholarships to France in the 1970s, and I was responsible for picking the 10 Engineering degrees from France which could be recognised by the government. Among the scholarship-holders in the first batch was Alan Chan, ex-CEO of SPH. In 1979, I was decorated with an Order by the French Government: the Chevalier des Palmes Academiques.

A rare chance to visit the Summer Residence of Chinese Premier at Bei Dai He (2001).

With Dr Richard Hu (extreme left, Aug 1990) and Chairman Michael Fam. Dr Hu, being the Finance Minister, was very supportive in funding the development of NTU. I count both men as key success factors in NTU.

With Bill Gates at the turn of the millennium (1999), a time when we were all worried about the Y2K bug, which turned out to be a false alarm. Technology has a way of surprising us.

Index

3-D printing 97
2G 184
3G 184
5G 184
98 Holdings 115

Abdullah Badawi 21
ACS 17, 21
Advanced Logic Research 132
aerodynamics 24, 25
aeronautics 91
AI 145, 147, 148, 150, 155
Alibaba 149
Amazon 146
Ambiavagar (Mr) 16
America 50
amphibious 135
Anderson, Marian 77
Ang, How Ghee 92
Ang, Kong Hwa 116
APJ 140
AR 146
architects 99
Arrau, Claudio 77
Artificial Intelligence 98, 144, 153
ASEAN 87, 107
ASEAN Federation of Engineering Organisations 86
Ash, Eric (Sir) 46, 128
Asia 140
Asian Association of Business Incubators Journal 141
A*STAR 96
ATM 98
auditor 99
Auditor General of Singapore 125
Augmented Reality 110
Australia 35, 104
Austrian bank 131
automation 93, 145, 147, 150, 153
automobile trade 133

Beh, Swan Gin 64
Bei Dai He 193
Beijing 102
Beijing University 55
Bellace, Janice 64
Benjamin, Brooke (Prof) 26
Bentley 132
Big Data 147, 150, 155
Bingei, Sendi 74
biomedical engineering 56
Biopolis 101
Bogaars, George 48, 125, 126, 178, 179
Brazil 116
Brenner, Sydney 92
British 140, 181
British Council 25
British government 28
British Naval Base 92
British Overseas Air Corporation 25
Broms, Bengt 41
Brown, John 45
Brussels 112
Bukit Timah 138
Bukit Tinggi Road 85
Business Excellence 113

CAAS 178
CAD-CAM 97
California 132
Cambridge 10, 15, 24–29, 31, 46, 83
Cambridge School Certificate 18
Cantonese 185
capital flight 187
Capitol Theatre 17, 18, 184
Carpay, Frans 97
Century tank 135
Chan, Alan 80, 193
Changi Prison 139
Changsha 133
Chengdu 133
Chen, Louis 164
Chen, Show Mao 74
Cheong, Pak Lo 13
Chevalier des Palmes Academiques 193
Chew, Choon Seng 32, 88
Chew, Keng Lian 169
Chief Justice 110
Chief of Defence Force 111
CHIJ 17, 18
China 1, 94, 99, 103, 107, 132, 133, 138, 141, 147, 189
Chinese 138
Chinese High School 60
Chng, Kai Jin 80
Choo, Seok Cheow 92
Choo, Thiam Siew 60
Choo, Winston 135
Circuit Breaker 189
City Cab 135
civil engineering 24
Civil Service 177
Clark, Brian 128
coal gasification 128
Cohen, Robert 72
ComfortDelgro 135
Commonwealth Engineering Council 41, 86
Commonwealth scholarship 24, 25
computing power 144
CONQUAS 99
contractors 99
Council for Professional and Technical Education 36, 40
Council of Engineering Institutions of South-East Asian Nation (CEISEAN) 85
CPTE Council 10
CSR 139
Curtis, Thomas 128

Dainton, Frederick (Lord) 46, 47
Darwin 49
D'Cruz, Raymond 13
Defence Science Organisation 96
Deming Prize 112
Dense Light 52
Department of Defence 95
Desker, Barry 101
developers 99
digital transformation 113
Dirac, Paul 28
disruptive innovation 143
disruptive technology 113, 153
DORSCON 188
Downtown line 138
drilling and production technology 128
Dubai 116
DVD 185

Eastern Pretech 115
East–West integrative medical treatments 75
Ecclestone, Bernice 123
EDB 100
EduTrust 104
e-learning 66, 67, 158
Electronic Road Pricing 107
email 184

embargo 185
Enabling Village 139
energy conservation 186
engineering 16, 19, 21, 22
 education 87
 excellence 87
England 20, 25, 45, 107
epidemic 188
EQA 112
ERP 2.0 107
ESG 113
Esplanade 80, 81
Europe 27, 100, 104, 138
European Foundation for Quality Management 112

Facebook 149, 184
Faculty of Engineering 84
Fam, Michael 44, 136, 139, 170–172, 175, 194
Faraday, Michael 154
FinTech 98
F&N 172
Fock, Siew Wah 131, 136, 139
Foodpanda 147
Ford 132, 173
Fourth Industrial Revolution 144
France 110, 193
Fusionopolis 101

Gates, Bill 194
GDP 185, 186
Gee Len 4, 72
Gek Khim 168
General Election 10
geopolitics 168
Germany 105, 110, 135, 140
GIC 111, 177
GINTIC 97
Goh, Chok Tong 10, 11, 45, 102
Goh, Keng Swee 10, 121
Goh, Yew Lin 80
Gongshang 7, 8
Google 149
Google Home 148
government 69
Government Investment Corporation 110
Grab 149
Grumman Aerospace 97
Gryphon Award 192

Harvard 53
Hawaii 92, 167
Hawking, Stephen 28
HDB 137, 170, 172
Heng, Swee Keat 63
Hewlett Packard 182
High Commission for Malaysia 176
higher education 56
Higher School Certificate (HSC) 19
high-powered business development VP 134
high speed aerodynamics 121
high technology 133
high wage policy 186
Ho Ching 35, 88
Hock Lee bus strike 20, 182
Ho, Kah Leong 59, 60
Ho, Kwon Ping 63, 64
Hong Kong 110, 120, 138, 141, 185, 188
Hon, Sui Sen 10, 48, 178
Hoong, Bee Lok 60
Housing and Development Board 110
Hsieh, Fu Hua 65
Huawei 184
Hunan 133
Hu, Richard 165, 171, 194
Hu, Robin 117

IDSS 176
IES/IEM Engineering Convention 85
IHL 108

Inderjit Singh 41
India 94, 107
Indian army 134
Indonesia 174, 175
industrial revolution 153, 154
Industry 4.0 144, 158
inflation 185
information and communication technology 154
Institute of Defence and Strategic Sciences 101
Institute of East Asia Philosophies 165
Institution of Engineers, the 45
integrated train testing centre 109
International Charitable Organisation 74
International Plaza 85
Israel 54, 185
Istana 161
Istomin, Eugene 77
IT 110, 155
Italy 189
ITE 99

Jaguar 132, 173
Jakarta 139
James 174, 175
Japan 92, 110, 140, 141, 146, 167
Japanese 1
Japanese MRT system 105
Japanese Occupation 2, 181
Japanese Union of Scientists and Engineers 113
Jobs, Steve 94
Jody, R 20
JTC 174
Jurong Island 101

Kampong Cham 95
Karawaci 175
Kazakhstan 49
Kent Ridge 186
Keppel shipyard 23
Khaw, Boon Wan 108
Knussen, Oliver 72
Koh, Boon Hui 88
Kuala Lumpur 91, 139
Kumagai Gumi 42, 172
Kwa, Chong Seng 32, 88

Lan Shui 78, 79
Leckie, James 128
Lee, Bee Wah 41
Lee, Boon Yang 119
Lee, Ek Tieng 88
Lee Foundation 180
Lee, Hsien Loong 78
Lee, Kong Chian 180
Lee, Kuan Yew 37, 43, 45, 48, 161, 167, 181, 182, 192
Lee Kuan Yew School of Public Policy 103
Lee, Moses 88
Lee, Seng Lip 88
Lee, Seng Tee 180
Lee, Seng Wee 174
Lee, Wei Ling (Dr) 33
Lee, Yock Suan 192
Lien, Ying Chow 131
Lim, Boon Heng 112
Lim, Chee Onn 127, 129
Lim, Chin Siong 7, 8
Lim, Chong Yah 37, 53
Lim, Chuan Poh 96
Lim, Kim San 110, 116
Lim, Leong Geok 137
Lim, Michael 105
Lim, Peter 2
Lim, Pin 40, 161
Lim, Raymond 74
Lim, Yew Hock 182
Lippo Group 174
Little India 108

Liu, Gretchen 175
lockdown 189
Lockheed 95
London 25, 76
London Olympic 140
London School of Economics 65
London Underground 138
Looi, Teik Soon 109
Low, Cecilia 61
Low, Kin Keong 60
LSE 66

M1 178
Maestro Choo Hoey 79
Magnolia Award 102
Malacca 133
Malaya 29
Malaysia 20, 22, 94, 133, 182
Malaysian 15
Malcolm Baldrige National Quality Award 112
Managerial Economics 103
Manchester 28
Manila 86
Mao Zedong 134
Marshall, David 181, 193
MAS 130, 170, 177
Massachusetts Institute of Technology 53
Mckinsey 61
MDIS 104
MINDEF 164
Minister for Communications and Information and the Arts 119
Minister for Defence 11
Minister for Trade and Industry 10, 112
Minister of the Trade Industry 37
Minister of Transport 108
Ministry of Defence 96
Ministry of Education 13, 37, 47, 67, 92, 187
Ministry of Finance 10, 34, 37–39, 47–49, 63, 110, 125, 178, 179
Ministry of National Development 99
Ministry of Trade and Industry 36, 51, 93, 100
missiles 96
MIT 54, 165
Mitsubishi 92
modular designs 100
MOE 65, 68
Monetary Authority of Singapore 62, 110
Montoya, Juan Pablo 124
mortar bomb 164
mother 14
Motorola 183
Mr and Mrs Lee Kuan Yew 33
MRT system 186
Munich 105
music 13, 14, 16, 71, 76, 77

Nagasaki 92
Nair, Devan 118
Nanyang Executive Centre 187
Nanyang Technopreneurship Centre 188
Nanyang University Alumni 130
Nanyang University Council 130
National Institute of Standards and Technology 112
national library 14
National Productivity Board 112, 113
National Science and Technology Board 51, 93
National Science Awards 94
national squad 15
National Wage Council 37
Neptune Orient line 10
Netflix 150
Net Investment Returns Contribution 111
New York 55, 97

Ng, Eng Hen 59
Ngiam, Tong Dow 49
NIE 191
Nobel Prize 92
Nokia 184
North-South line 136
North Vietnam 95
NTUC 37, 118, 146
nuclear energy 129

Ocean Engineering Centre 167
Ocean Tower 125, 126, 178
Odeon Theatre 17, 18, 184
offshore wind 128
oil rigs 179
Oman 152
one-north 101
Ong, Beng Seng 115, 123
Ong, Teng Cheong 176, 192
OPEC 185
Opel 135
Oppo 184
Osaka 141
OUB 130, 131
OUE 131
Outram Park 85

Pacific 140
Pantai Valley 21
PAP 20, 182
Paris 48, 49
Partek 115
Pasir Panjang English School 8
Pasir Panjang Primary School 9
Patel, Minoo 128
PCB 133
Pearl's Hill 96
Peking U 102
People Action Party 181
Persian Gulf 151
petrochemical industry 101
Pho, Robert 75
Phua, KK 60
Pillay, JY 80, 88, 110
Poh, Choon Ann 60
polytechnic 31, 32
Port Dickson 23
potentially disruptive
technologies 128
practice-oriented education 43
precast components 100
prefabrication 100
President Scholarship 20
Pricewaterhouse 125
Prime Minister 11
Private Equity Company 134
PSC 110
PSLE 8
Public Transport Council 138
PWC 99

Qaboos Bin Said 151
Quality Class Certifications 113
Queen, the 25

Radio Shack 183
Raffles City 17
Raffles Country Club 109
Raffles Girls School 18
Raffles Marina 116
Raffles Place 125
Rahman, G 20
R&D 106, 134
Reagan administration 112
Real Time Accounting 140
recession 138
Rediffusion 185
REIT 119
Republic Polytechnic 57, 58
RFID chips 146
RGS 18, 61, 62
Rhodes, Cecil John 73
Riady, Mochtar 174, 175
Riady, Stephen 80, 131

Ricci, Ruggiero 78
Robinson 174
robotic engineering 155
robotic surgery 150
Rolls-Royce 132, 173
Royal School of Music 72
Royal Society of Britain 93

Saint Joseph 17
Sayyid Asaad 152
School 9
Schumacher, Ralf 124
Schuster, Wolfgang 75
Science and technology 97
Scotland 27
seaport 167
Second World War 1
Sembawang Group 116
SGX 139
shale oil 129
Shanghai 101, 133
Shanghai Municipal 102
Sharples, Malcolm 128
Shell 171
Shell refinery 23
Shenyang 133
Shenzhen 133
SIA 172
Silicon Valley 71
SIM 59, 64, 65, 69–71
Sim, Kee Boon 126, 127, 166, 178, 179
SIMTech 97
Singapore 16
Singaporeans 24
Singapore China Foundation 102
Singapore Government 24
Singapore Institute of Management 62
Singapore Management University 65
Singapore Polytechnic 56
Singapore Science Park 101
Singapore Stock Exchange 139
sister 14
SIT 59
SJI 21
Sloan School 165
sms 184
SMU 177
social distancing 189
Soon, Freddy 112
South Korea 72, 110, 141, 146
South Korean 138
sports 12
Sri Lanka 35
SR Nathan 68, 101, 110, 175, 190
SSO Council 175
St Andrew 17
state scholarship 20
St Catharine's College 26, 47
ST Kinetics 134
Straits Times 2, 118
sub-sea related developments 129
Sudbin, Yevgeny 72
Sultanate of Oman 151
SUSS 59, 68, 95, 107
SUTD 59, 175
Suzhou 133

Taiwan 99, 138, 141
Tambyah, Richard 13
Tam, James 55
Tan, Bernard 92
Tan, Boon Teik 78
Tan, Cheng Bock 105
Tan, Chin Tuan 167–169
Tan, Chok Kian 63
Tan, Eng Chye 34
Tan, Eng Liang 74
Tange, Kenzo 39, 42, 49
Tang, IF 100, 132, 167, 173, 174
Tan, Jin Kwee 20
Tan, Lay Pheng 57

Tan, Mary 169
Tan, Tony 20, 38, 44, 47, 48, 62, 63, 66, 119, 165, 169, 171, 176
Tata Steel 116
Tat Jen 4
Tay, Eng Soon 19
Tay, Sin Yan 84
telephone communication 183
Temasek Holdings 111, 136, 137, 177
Tengku Abdul Rahman 182
Teo, Chee Hean 59, 60
Teo, Siok Guan 3
Terminal 5 140
Thailand 116, 164, 187
Tharman Shanmugaratnam 61, 62, 68
Think Tank 176
Thomson East Coast line 138
Tiong Bahru 1, 9, 14
Titan 28, 183
TODAY 121
Toh, Chin Chye 86, 91, 166, 186
Tokyo 112
Total Quality Management 112
Tote Board 165
transmission 188
Twitter 149, 184

UK 27, 35, 36, 104, 107, 110, 138, 166, 189
Ullring, Sven Bang 128
UNDP 173
United Engineers 133
United Overseas Land 131
United States 95, 189
United States Secretary of Commerce 112
Universitas Pelita Harapan 175
University 19
University of London 65
University of Loughborough 191
University of Malaya 20, 21, 22, 31, 83, 91
University of Oxford 73
University of Singapore 139
University of Surrey 49
University of Washington 54
URA 64
US 35, 36, 104, 107, 112, 138, 163

vehicle 135
Victoria Concert Hall 80, 81
Victoria Memorial Hall 76
Vietnam War 95
Vij, Kirpa Ram 96
Vivo 184
voice communication 184
Volvo 173
VR 110

wafer fabrication 93
Wales 27
Wall Street Journal 133
Warwick University 54
Washington 105
waste-to-energy 128
WeChat 149, 184
Wee, Cho Yaw 42, 121, 130, 165, 166, 169
Wee, Ee Cheong 131
Wee, Kim Wee 43, 190
Wen Jiabao 11, 102
Wharton 64, 180
WhatsApp 184
wide-scale automation 143
wife 3, 18
Wi-Fi 187
Wong, Aline 92
Wong, Nik Liong 88
Wuyi San 75

Xiaomi 134, 184

Y2K bug 194
YBA 101
Yeo, Choo Kok 7
Yeo, George 42, 80
Yeo, Li Pheow 57
Yeo, Ning Hong 128, 135, 139
Yeo, Philip 96
Yip, John 63
Young, John 16
YouTube 150
Yu, Chun Yee 59, 77

Zurich 188